GURWITSCH'S RELEVANCY FOR COGNITIVE SCIENCE

CONTRIBUTIONS TO PHENOMENOLOGY

IN COOPERATION WITH THE CENTER FOR ADVANCED RESEARCH IN PHENOMENOLOGY

Volume 52

Scope

The purpose of this series is to foster the development of phenomenological philosophy through creative research. Contemporary issues in philosophy, other disciplines and in culture generally, offer opportunities for the application of phenomenological methods that call for creative responses. Although the work of several generations of thinkers has provided phenomenology with many results with which to approach these challenges, a truly successful response to them will require building on this work with new analyses and methodological innovations.

GURWITSCH'S RELEVANCY FOR COGNITIVE SCIENCE

Edited by

LESTER EMBREE

Florida Atlantic University,
Boca Raton, FL, U.S.A.

Springer

A C.I.P. Catalogue record for this book is available from the Library of Congress.

ISBN 1-4020-2891-1 (HB)
ISBN 1-4020-2892-X (e-book)

Published by Springer,
P.O. Box 17, 3300 AA Dordrecht, The Netherlands.

Sold and distributed in North, Central and South America
by Springer,
101 Philip Drive, Norwell, MA 02061, U.S.A.

In all other countries, sold and distributed
by Springer,
P.O. Box 322, 3300 AH Dordrecht, The Netherlands.

Printed on acid-free paper

springeronline.com

Printed in the Netherlands.

Table of Contents

Preface

When I heard the rumor that the findings about the central nervous system obtained with new technology, such as Magnetic Resonance Imaging (MRI) and Positron Emission Tomography (PET), were too subtle to correlate with the crude results of many decades of behavioristic psychology, and that some psychologists were now turning to descriptions of subjective phenomena in William James, Edmund Husserl, and Maurice Merleau-Ponty—and even in Buddhism—I asked myself, "Why not Aron Gurwitsch as well?" After all, my teacher regularly reflected on the types, basic concepts, and methods of psychology, worked with Adhémar Gelb and Kurt Goldstein in the institute investigating brain-injured veterans at Frankfurt in the 1920s, conspicuously employed Gestalt theory to revise central Husserlian doctrines, and taught Merleau-Ponty a thing or two. That the last book from his Nachlass had recently been published and that I had recently written an essay on his theory of psychology no doubt helped crystallize this project for me.[1]

What is "cognitive science"? At one point in assembling this volume I polled the participants, asking whether they preferred "the cognitive sciences" or "cognitive science." Most who answered preferred the latter expression. There is still some vagueness here for me, but I do suspect that cognitive science is another example of what I call a "multidiscipline."[2] A multidiscipline includes participants who confront a set of issues that is best approached under more than one disciplinary perspective. There is little or no merger of disciplinary perspectives involved, but there is considerable will to learn from work in other fields. Colleagues also told me that the term "cognitive science" took root in the 1970s, following a Slone Foundation grant for conferences, and that The Cognitive Science Society asserts that it "brings together researchers from many fields with a common goal: Understanding the human mind. We promote scientific interchange across disciplines, including Artificial Intelligence, Linguistics, Anthropology, Psychology, Neuroscience, Philosophy, and Education." The philosophers involved then were originally in the analytic tradition, but of late also include some in the phenomenological tradition.[3]

---

[1] Aron Gurwitsch, *Esquisse de la phénoménologie constitutive*, ed. José Huertas-Jourda, introduction by Lester Embree (Paris: J. Vrin, 2002) and Lester Embree, "Gurwitsch's Theory of Cultural-Scientific Phenomenological Psychology," *Husserl Studies* 19 (2003): 43-70.

[2] Cf. Lester Embree and Stanford Lyman, "Ethnic Studies as Multi-Discipline and Phenomenology," in *Phenomenology of the Cultural Disciplines*, ed. Mano Daniel and Lester Embree (Dordrecht: Kluwer Academic Publishers, 1994), 211-49.

[3] Cf. Jean Petitot, Francisco J. Varela, Bernard Pachoud, and Jean-Michel Roy, *Naturalizing Phenomenology: Issues in Contemporary Phenomenology and Cognitive* Science (Stanford: Stanford University Press, 1999) and the new journal, *Phenomenology and the Cognitive Sciences.*

With the goal of showing that Gurwitsch is relevant for cognitive science, it only remained to identify a set of suitable colleagues and to bring them together for another of the research symposia I had been holding at the Seagate Hotel and Beach Club in Delray Beach, Florida since 1991. Dr. Olav Wiegand proved invaluable in this connection, contributed well to the discussions, and I regret that he needed to withdraw from the project. I have depended on Dr. Kirk Besmer, the William F. Dietrich Fellow in Philosophy at Florida Atlantic University in 2001-2003, and his successor, Mr. Daniel Marcelle in 2003-2005 for help in ways too numerous to list.

Lester Embree

Delray Beach
November 2003

Introduction:

Explanation, the Noematic Core, the Essays, and More Relevancies for Cognitive Science

Lester Embree
Florida Atlantic University

Abstract

Before some commentary on the contents of this volume, something is offered about Aron Gurwitsch's model of explanation, which involves the central nervous system, and how, for him, the noematic core, which is fundamental to his phenomenological descriptions, can be abstractly discerned.

I. Psychological Explanation

Like that of his master Edmund Husserl, Aron Gurwitsch's philosophical project can be characterized as a philosophy of science, or better, a theory of science—*Wissenschaftslehre* or *Wissenschaftstheorie* in German. Many metaphysically preoccupied post-Husserlian phenomenologists have forgotten this original philosophical project of our tradition. My teacher was delighted when I suggested the title for his last collection of essays.[1] His "science theory," as *Wissenschaftslehre* can also be rendered, includes reflections not only on logic and mathematics, physics, biology, and the cultural sciences, but also on psychology. Among other things in his theory of psychology, he supports a revision of the traditional model of explanation.

It is my impression that what cognitive scientists (and analytic philosophers) most find lacking in phenomenology is a preoccupation with causal explanation. The downplaying of causality is quite deliberate in phenomenological philosophy, but Gurwitsch does not extend it to psychology as a positive science. Thus in his masterpiece he writes as follows:

> From the psychological point of view, acts of consciousness are considered as events of a specific nature occurring in the same real world and in the same objective time as events of other kinds, e.g., organismic processes (especially of the nervous system), and physical processes of all sorts. Acts of con-

[1] Aron Gurwitsch, *Phenomenology and the Theory of Science,* ed. Lester Embree (Evanston, IL: Northwestern University Press, 1974), hereafter cited as "PTS."

L. Embree (ed.), Gurwitsch's Relevancy for Cognitive Science, 1-24.

> sciousness are causally or functionally related to physiological processes which, in turn, are aroused by certain physical processes, namely, external stimulations of the sense-organs. It is with regard to such functional dependencies that consciousness may be, and actually is, studied in psychology. Psychology thus is a positive science. Like the other positive sciences, psychology chooses a well circumscribed realm of reality, as one mundane realm among others, and connected with these realms. It is in full conformity with this choice, that, in its exploration and explanation of consciousness, psychology continues and, partly, relies upon the physical and biological sciences.[2]

Following Gestaltists of the Berlin School, Gurwitsch shows how the constancy hypothesis played a fateful role in pre-Gestaltist psychological explanation. Briefly, what is hypothesized is a one-to-one correlation between external stimuli and the mental events usually called sensations, sensa, or sense data (on Gurwitsch's interpretation, Husserl's hyletic data are sense data). Interestingly, however, subjects often do not perceive what the stimulation of their sense organs lead psychologists to predict that they will perceive. The traditional solutions to this problem assume dualistic conceptions of sensuous perception in which, in addition to the sense data, secondary processes are posited. Gurwitsch focuses on the work of Alexis Meinong, Vittorio Benussi, and the School of Graz (FC, Part I, Ch. 3), but philosophers today are probably most familiar with this model through the equation whereby

perception = sensation + interpretation.

Gestalt psychologists call this the "meaning theory" or the "interpretation theory" (FC, 44 n. 46). This theory is attractive because the recourse to sensations can support the belief that physical reality is always contacted, while all sorts of differences, including those of class, ethnicity, gender, and generation, can be attributed to interpretation, interpretive frameworks, etc., and philosophers as well as scientists can then analyze these added conceptual components in perception, looking for them first of all in language.

There are two difficulties with this traditional model. To begin with, it is unacceptable because, as Wolfgang Köhler already showed in 1913, it is "nei-

[2] Aron Gurwitsch, *The Field of Consciousness* (Pittsburgh: Duquesne University Press, 1964), 157-58. Hereafter cited as "FC." Cf. Aron Gurwitsch, *Studies in Phenomenology and Psychology* (Evanston, IL: Northwestern University Press, 1966), 89, hereafter cited as "SPP," and Aron Gurwitsch, *Marginal Consciousness*, ed. Lester Embree (Athens, OH: Ohio University Press, 1985), 15-16, hereafter cited as "MC."

ther self-evident nor verifiable" (FC, 91). Moreover, if one actually reflects on things-as-perceived, no phenomenological difference can be discerned between the sense data and a meaning, interpretation, or other subjective addition. "While listening to a melody or seeing a geometrical configuration, one is unaware of any non-sensory process, nor is the theme experienced as consisting of two strata of different origin. On the contrary, the melody or the configuration, as given in immediate experience, appear as both homogeneous and altogether a matter of sensibility" (FC, 88-89).

I hasten to add that Gurwitsch was well aware of a discernable difference between the "meaning" and the "meaning carrier" in language (see FC, 267 ff.; PTS, 228 ff.), and acknowledges that "Husserl admits a certain parallelism between the perceptual apprehension of material things and the comprehension of meanings in the proper and more narrow sense ..." (FC, 269). What is at issue, however, is precisely the perceptual apprehension of material things when one listens to their sounds or looks at their shapes. It is the interpretationist theory of perception, not that of language, that is rejected.

To introduce the alternative he advocates, Gurwitsch contrasts two models of explanation that he formulates in mathematical terms:

> Denoting the stimuli and external conditions playing a part by x_e, the subjective conditions by x_i (internal conditions), the resulting percept by P, we may illustrate Benussi's theory by the mathematical expression: $P = f_1 (x_e) + f_2 (x_i)$, f_1 and f_2 standing for functional dependencies. This expression renders Benussi's interpretation of the percept as consisting of two heterogeneous strata, one of which depends only upon external conditions, and the other, founded by the former, only upon internal conditions. If the external conditions are kept constant, while allowing the internal conditions to vary, the illustrating mathematical expression assumes the form: $P = \text{Const.} + f_2 (x_i)$. This fits very well with Benussi's view in which the percepts have a constant common stratum, but differ as to the variable part which, on account of its origin, is the 'higher' stratum.
>
> To expound the Gestalt theoretical interpretation ..., let us consider, for the sake of simplicity, the case of mental isolation only. We compare two perceptions: P_1 when the subject looks at the figure as a whole; P_2 when he concentrates on certain lines, disregarding others. With respect to the two perceptions, one must insist upon the identity of the stimuli, and upon the difference in the attitude of the perceiving subject. When in consequence of a change in attitude, different percepts result, the only warranted conclusion is that *the percept*

> *depends upon both external and internal conditions.* Abandoning the constancy-hypothesis, one can no longer justify distinguishing between two strata within the percept, either stratum varying in dependence upon conditions of only one special kind ... *if the percept varies in dependence upon both external and internal conditions, it does so as a whole and as a homogeneous entity.* (FC, 94-95)
>
> A mathematical illustration of the Gestalt theoretical view is provided by the expression $P = f(x_e, x_i)$. If the external conditions are kept constant, this expression becomes $P = f$ (constant, x_i), which obviously is not the same thing as $P =$ constant $+ f_2(x_i)$. Especially as to the above-mentioned percepts $P_1 = f(x_e, x_i)$ and $P_2 = f(x_e, x'_i)$—the external conditions being kept constant, while the internal conditions are permitted to vary—Gestalt theory maintains that P_1 and P_2 differ really and substantially from one another and rejects the interpretation of P_2 as somehow "contained" in P_1. (FC, 95)

Here the internal conditions mentioned are attitudes (attitudes will be discussed below more extensively), but they are not the only internal conditions.

> Internal conditions comprise any operational factors other than the actual stimulation of sense-organs. More, therefore, than the adopted attitude of the experiencing subject must be taken into consideration, when one time, he looks at a presented figure as a whole, while another time he tries to single out one or another of the component lines. Among the internal conditions, the past of the individual is of paramount importance. (FC, 96)

Concerning past experience, traditional theory has memories, images, etc., that somehow come back from the past and are added to current sense data. "If the suit of polished armor 'looks' cold, hard, and smooth, it is because sensations of coldness, hardness, and smoothness have been experienced in the past and the sight of the object 'instantly reinstates and steadily maintains' those sensations" (FC, 250). But no such data can be observed returning from the past to attach to present sense data. Instead, what one actually does perceive due to past experience can be described as follows:

> What in the visual appearance is given in direct sense experience is but one constituent of the total percept and as such is qualified by, and organized with respect to, other constituents.

> Seeing an orange, we perceive the spherical orange surface as enveloping juicy and tasty content. In the visual perception of a piece of polished iron, a grayish color appears as covering a hard, cool, and smooth surface of a certain geometric shape. The piece of iron does look hard, cool, and smooth because its visual appearance is essentially determined by references it implies to tactile properties not given in direct sense-experience. (FC, 278)

In other words, what is directly perceived predelineates certain possible future perceptions, present references to perceivings and appearances in the future being not the same as data returning from the past.

Finally, where past experience is concerned, it is not necessarily a matter of remembering what was previously perceived, which is of course also often possible. Concerning what is necessary in this connection, however, Gurwitsch again follows the Gestaltists in holding that "previous experiences leave 'traces' in the nervous system which modify that system as the medium in which processes aroused by external stimulation take place" (FC, 98). This may suffice to show that the model of explanation that Gurwitsch supports recognizes not only external stimuli, but also internal conditions that pertain to such things as attitudes and modifications of the central nervous system by experience.

With his acceptance of the Gestaltist revision of the explanatory model in psychology, Gurwitsch was able to appreciate the descriptive findings of Gestalt theory and use them to revise Husserl's phenomenology. And where methodology is concerned, he considers the abandonment of the constancy hypothesis an "*incipient* phenomenological reduction" (FC, 168). This is not the occasion to discuss whether what is involved in this case is the phenomenological-psychological or the transcendental-philosophical *epochē,* reduction, and purification.

It deserves mention, however, that descriptions produced in transcendental-philosophical phenomenology can be interpreted as mundane or worldly. This was Husserl's assertion in his "Nachwort zu meinen 'Ideen…'" (1930), which text Gurwitsch reviewed in 1932, and his close friend Alfred Schutz relied upon this mundanizing interpretation for his own investigations in the "constitutive phenomenology of the natural attitude" (Husserl's phrase). Moreover, in his *magnum opus* (completed in 1953), Gurwitsch himself recognized the validity of Schutz's "phenomenological psychology" carried out within the natural attitude (FC, 400-401), and in 1966 he also studied Husserl's own 1925 lectures on phenomenological psychology thoroughly and critically (PTS, 77-112). In speaking of phenomenology, then, one needs explicitly or by context to qualify that title as philosophical or psychological. I was interested to see that the colleagues at our research symposium did not refer to this distinction, or to the interpretive move in question, but seemed simply to take it for granted.

II. About Some Terminology

What contrasts with explanation is description, which is concerned not with why something is, but with what it is. It would seem obvious that explanation presupposes description, i.e., that the first need is for description. With a description in hand, one knows what to explain and can try to do so. Common sense and ordinary language may often seem to be adequate for descriptive and explanatory purposes. But this is dubious, if only because the psychological terminology found in ordinary language is so often a sediment of earlier psychological theory. For example, it is doubtful that concern with unconscious psychological motivation goes back in the history of common sense further than Sigmund Freud. This is, of course, part of a wider problem: not only does science shape culture, but even obsolete science can continue to exert an influence.

Wherever its terms have come from, ordinary language can mislead. For example, the centermost concept in phenomenology is traditionally expressed with the word "intentionality" and then there is talk of "intentional experiences" or "intentional acts." Now in ordinary English intentions and intentionality are concerned with purposes and plans for the future, often with moral connotations. According to phenomenology, however, the remembering of earlier events and the thinking of atemporal things are also characterized by intentionality and are thus also intentional processes, as is perceiving. Confusions can be resisted by employing technical terminology in this connection.

Having heard "intentionality" used at our symposium in a seemingly phenomenological context but with the ordinary English signification, I mentioned that my other teacher, Dorion Cairns, had revived the verb "to intend to," current in Isaac Newton's time, and derived the adjective "intentive" and the abstract noun "intentiveness." While these words have general signification, there are times when "purposiveness," "purposing," etc., are best for the specific intentiveness to the practical future. I doubt that any who heard me at the symposium were moved thereupon to reform their linguistic habits, so I have seized this opportunity to repeat the comment in print! The more general point is that phenomenology, whether philosophical or psychological, needs constantly to work at clarifying and refining its technical language. More efforts of this sort will be found below.

III. Reflective or Noematic Analysis of Looking at a Building

The following exposition of what is central to Gurwitsch's work is more systematic than what is to be found in any one place in his oeuvre, and some of the terminology is refined. Quotations will nevertheless show that he accepts most of the distinctions made. The chief source he relies on is Husserl's *Ideen I*

(1913) and the main example he employs is that of *looking at a building,* which one can engage in seriously in perceiving or remembering or feign in perception or pretend to remember. Such a case actually involves a great deal. For example, the following passage reflects the difference between the theme and the rest of the field of consciousness.

> In looking at a house and choosing it as the theme of our present perception, we perceive the house as flanked by other buildings, as located beside a park. We see the street on which the house is located, the passersby, and the cars moving along. The other buildings on the street, the people, the cars, the park, and all other things perceived, while we are dealing with the perceived house as our theme, do not, of course, become themes themselves. Rather they form a background around the house, the theme of our present perception. (FC, 320-21)

Ten steps lead to the heart of Gurwitsch's analysis of such a case. The first step of this analysis is to recognize that what one looks at in the example is the whole building.

> To perceive a building from a certain side is not to see that side as in a drawing, or as a wing on a stage, or a silhouette. The building itself is given through any single perception, though it appears from a particular side. Every single perception proves to be a perception of the thing itself and as a whole, though from a determinate point of view. To express this in phenomenological terms, the experience of any single appearance proves tantamount to the apprehension of the entire noematic system through the apprehension of one of its members. It is by virtue of such references to further noemata that the appearance presently actualized is essentially determined as appearance of the perceived thing itself. (FC, 219; cf. PTS, 253)

This statement seems confined to the visual, but a passage quoted above in relation to explanation includes gustatory and tactual appearances and such matters will be mentioned in later steps of the analysis.

Second, we can recognize that there can be an episode of looking at the building in which the appearances change:

> If we are interested in a building, in order to see it from many sides and angles we walk around it, come close to it, draw away from it. In the course of our perceptual exploration, we

> are presented with a multiplicity of perceptual appearances which succeed each other more or less immediately. All these appearances, while differing to a greater or lesser extent, are experienced as perceptual presentations of one and the same thing, namely, the building in which we are interested. (FC, 2; cf. MC, 32, MC, 53)

We can, of course, be conscious also of the building from inside:

> Sitting in a room, we are aware of the things behind our back through horizonal consciousness as things which fit into a room of this kind, whether or not these things present themselves with full determinacy regarding their particular details. In this case, the experience of context rests upon the experienced suitability of the things in question to the purpose of the room, the awareness of this purpose being conveyed by those things which are given in actual perception. (MC, 47; cf. FC, 340)

Now we can recognize more clearly that there is a difference between the thing that is seen and how it appears, i.e., how it is seen. This is a relation between the same building and the multiplicity of what were just called "appearances" (although other names are used as well).

In the third place, the present analysis will be confined to looking at a building from a fixed place outside. Then, "while abiding by the same point of observation, we alternately open and close our eyes, or ..., after an absence from a certain point of observation, we return to it—so that the same noema corresponds to a multiplicity of acts" (PTS, 247). Thus there is again a multiplicity in contrast with an identity. In short, the appearance, now called "noema," is located in relation to the thing looked at, on the one hand, and in relation to the multiple seeings or lookings at the building, on the other hand.

Fourth, the analysis of the noema in the case of a looking at a building has an aspect that raises terminological questions. More than sensuous perception strictly speaking is involved:

> What we encounter are cultural objects, objects of value, e.g., works of art, [and] buildings which serve specific purposes, like abodes, places for work, schools, libraries, churches, and so on. Objects pertaining to the life-world present themselves as tools, instruments, and utensils related to human needs and desires; they have to be handled and used in appropriate ways to satisfy those needs and to yield desired results. It is the spe-

> cific sense of their instrumentality which essentially defines those objects and makes them be what they are, that is, what they mean to the members of the sociohistorical group to whose life-world they belong. (PTS, 143-44)

A terminological refinement may help in this connection. The case of "looking at a building" emphasizes visual perception, but besides being given in that manner, buildings also have uses. In this connection, Gurwitsch speaks of "instrumentality," which is a type of "functional character" (FC, 39 ff., FC, 98 ff.). A broader concept that includes the uses of things as well as strictly visual percepts or sights (and much more besides) can be expressed with words based on the verb "to encounter," even though this actually *reverses* how Gurwitsch, who was following Robert Sokolowski in this respect, began to use the word "encounter" in his last writings (e.g., PTS, 139 n.). Alternative expressions to Husserl's "noesis" and "noema" can then be "encountering" and "thing-as-encountered." Consequently, the thing-as-seen is a part of thing-as-encountered and is the result of abstracting from uses and the like.

If there is a difference between the house as seen and the uses that it has, what is it that noetically correlates with uses and parallels the seeing component within the encountering? In a broad signification that includes inactual using, e.g., inclinations to use, and habitual forms, the parallel to seeing can be called "willing." Seeing and willing can then be said to be prominent components of encounterings in which a building is encountered and different from but intentively correlative to the building-as-seen and the building-as-willed within the building-as-encountered. (The use of hyphens in this way is derived from James's *Principles of Psychology*.)[3] Further noetic and noematic components of encountering a building will be distinguished presently.

In the fifth place, to recognize the noetic and noematic components just mentioned, one needs to reflect. Reflecting is a different sort of encountering than the sort of encountering that includes seeing and willing as just described. More specifically, it contrasts with the straightforward encountering in which all humans, even phenomenologists, spend most of their waking lives (it is not easy to say whether members of other biological species ever reflect). Reflecting is clearly not sensuous, but it is still an observing. "Here as everywhere reflection consists in nothing other than rendering explicit and disengaging what had pertained to the act reflected upon, albeit in an undisclosed fashion, previously to

---

[3] When I sent out the first draft of the present introduction to the colleagues in the project, Gallagher responded by saying "There is a good parallel here to contemporary neuroscientific theory of vision as involving two distinct but interrelated pathways—the ventral pathway that allows for object recognition and the dorsal pathway that guides motor behavior for, e.g., instrumental manipulation (Milner, A.D. & Goodale, M.A. (1995) *The Visual Brain in Action*, Oxford: Oxford University Press.)"

and independently of its being reflected upon."[4] And reflection takes at least two forms, i.e., reflective observing of the encountering and reflective observing of the thing-as-encountered (whether there is a reflectively observable ego or, better "I" who encounters, which Gurwitsch disputes, does not need to be decided here).

As I have explicated elsewhere, Aron Gurwitsch was a noematically oriented constitutive phenomenologist.[5] Briefly, constitutive phenomenology is the project of accounting for things that are or can be encountered in terms of the actual or possible encounterings of them—or as Gurwitsch would say, in terms of consciousness. His emphasis is on the thing-as-encountered, in this case a building. And reflection discloses that a thing is not only given in some way or other, but also posited in various ways.

In the sixth place, there is more mentioned by Gurwitsch that relates to positing and the posited, i.e., the positional or thetic, than is obvious in his writings because the pertinent passages are few and scattered. How a building is volitionally posited has already been described somewhat. To say more, it might be added that a house can be encountered as able to protect persons and property. It can thus be used against the weather and thieves, but it can also be used for privacy, it can be used as an investment, it can be used to show accomplishment and claim social status, etc., etc. Reflective analysis can hence classify uses as negative ("against") and positive ("for"), and also as intrinsic, i.e., as characteristic of purposes or ends, and extrinsic, i.e., as characteristic of means.[6]

Yet reflection easily discloses that a thing-as-encountered is posited in at least two other ways that correlate respectively with valuing and believing. The valuing component in the building-encountering might be an aesthetic liking, and correlative to that, the building would have positive aesthetic value and

---

[4]PTS, 243; cf. SPP, 272 ff. See also Lester Embree, "Gestalt Law in Phenomenological Perspective," *Journal of Phenomenological Psychology* 10 (1979): 112-27.

[5] See Lester Embree, "Editor's Introduction: Gurwitsch's Phenomenology of the Margin, Body, and Being," in MC, xi-xlii."

[6]The emphasis in Gurwitsch regarding the positional is on the volitional. He recognizes willing, preferring, or deciding; cf. Aron Gurwitsch, *Human Encounters in the Social World*, ed. Alexandre Métraux, trans. Fred Kersten (Pittsburgh: Duquesne University Press, 1979), 45, hereafter cited as "HE." Moreover, conscious acts of all types have their noemata (EPC, 216), and thus there is a noema of willing (SPP, 339). His concern is especially with what can be called secondarily passive or habitual willing. In most of his descriptions, objects have correlative "functional characters" (FC, 49) and thereby are functional objects, but they can also be called "use objects" and said to have "uses." See Lester Embree, "Some Noetico-Noematic Analyses of Action and Practical Life," in *The Phenomenology of the Noema,* ed. John Drummond and Lester Embree (Dordrecht: Kluwer Academic Publishers, 1992), 157-210. Gurwitsch does not distinguish positive and negative modalities of willing and the corresponding praxic characteristics in the noema, but he does refer to the difference between ends and means (FC, 22; EPC, 322), and thus implicitly to that between end uses and means uses.

could be said to be beautiful, pretty, or at least attractive. Or it can be ugly correlative with disliking. And one can be apathetic about it, which is a third valuational modality, and in this case it would have neutral value. Since there are already interesting parallels between the valuational and the volitional, one might then be led to wonder if there is volitional neutrality. That there are intrinsic and extrinsic values parallel to end uses and means uses is relatively obvious.[7]

The type of positing (Husserl and Husserlians also speak of "thesis" and the "thetic" in this connection) that is called *doxa* or belief is also relatively parallel or analogous in most respects. (Yet, in English, one "believes" propositions but "believes in" things.) There can be positive believing in or negative disbelieving in the thing, e.g., the building, and an attitude of so-called suspended judgment or skeptical neutrality can also be adopted toward it. Correlatively, one can reflectively discern in the thing-as-encountered not only uses and values of various sorts, but also belief characters. Whether there are intrinsic and extrinsic belief characters is another question that does not need to be answered on this occasion.

When belief is firm, this can be termed positive or negative certitude, but positive and negative believing and things-as-believed-in (or as-disbelieved-in) can be shaky instead of firm, in which case terms such as conjecture or probability can be used. The firm/shaky distinction also holds for the valuational and volitional, in the latter case being expressed in terms of resoluteness and hesitancy. Husserl emphasized doxic modalities, but Gurwitsch also summarizes that account in passing, e.g., "a certain state of affairs (the identical matter of a proposition) is asserted or denied, doubted, questioned, or deemed probable."[8]

This brief analysis should suffice to show that things-as-encountered are posited in many ways. Gurwitsch alludes to practically all that has just been expounded, but his results indicate that most of the time he tacitly disregards or abstracts from how things are posited in our encounterings of them.

But these too can nevertheless be distinguished reflectively:

> When perceiving a hammer as a tool, we may adopt an attitude of reflection and proceed to an analysis of the perception and the corresponding perceptual noema. We disengage the

---

[7] Regarding valuing, Gurwitsch does discuss synthetic and synthesized loving (SPP, 245), mentions values (HE, 66, HE, 173 n. 96; EPC, 114, f., EPC, 121, EPC, 165), and correlates acts of appreciation with the values conferred on the object (EPC, 127), but he does not distinguish negative and positive modalities, species of valuing and values, or intrinsic and extrinsic values, as Husserl does. See Lester Embree, "Advances Regarding Valuation and Action in Husserl's *Ideas II*," in *Issues in Husserl's "Ideas II,"* ed. Thomas Nenon and Lester Embree (Dordrecht: Kluwer Academic Publishers, 1996), 173-98.

[8] FC, 327; cf. EPC, 127: *"Si une act est vécu dans les modalités de la présomption, de la conjecture, de la question, du doute, l'objet noématique correspondant porte respectivement les caractères 'possible,' 'probable,' 'problématique,' 'douteux.'"*

> noematic constituents and bring them out in the roles they play for the constitution of what presents itself through the act of perception, considered exactly as it appears through the given act. Among such constituents, we mention functional characters upon which depends the specific instrumentality of the hammer perceived as a tool. (FC, 235)

Presumably Gurwitsch would also say that values and belief characters can be disengaged in the same way.[9]

Seventh, things are not only posited but also given, so that there is also discernment and classification of "manners of givenness" *(Gegebenheitsweise)* in the thing-as-encountered. Gurwitsch speaks of how things are given "in perception, in memory, in clear intuition (*klar anschaulich*), in thought *(denkmässig),* etc." (PTS, 248). Correlatively, for temporal things, there are noematic characters of being present: being past or having been present; and of going to be present (EPC, 136; cf. SPP, 197). Thus there are indeed remarks about memory and imagination, but he is chiefly concerned with perception. Above all, there is the character of "perceptivity" (FC, 179). And there are further types of givenness: "The thing presents itself in a certain orientation relative to the observer and his standpoint as near or far, as at the center of the visual field or more toward the periphery, etc. Finally, the manner of appearance of the perceived thing varies according to whether it is seen in bright daylight or at dusk, in a fog, and so forth" (SPP, 332).

In the eighth place, the running example for this analysis is looking at a building, but what is looking? More adequately it can be called a type of visual perceiving. The visual can be distinguished from the tactual, auditory, olfactory, and gustatory. Houses have their smells (at least inside), the walls are smooth or rough to the touch, one hears something distinctive when the door is knocked on or the wind rattles the windows, and so on. Gurwitsch's analyses are typically visualistic, yet it deserves mention that he not only alludes to other sensuous modalities, but refers to audition on a number of occasions (e.g., FC, 123, 137, 233, 258). He also accepted from the Gestaltists what can be called the doctrine of synaesthesia. It has already been pointed out that a suit of polished armor "looks" cold, hard, and smooth and that an orange "looks" tasty; presumably one

---

[9] The difference between the naturalistic and the cultural sciences also depends on this abstraction. Cf. Lester Embree, "A Gurwitschean Model of Culture or How to Use a Spear Thrower," in *To Work at the Foundations: Essays on the Phenomenology of Aron Gurwitsch*, ed. J. Claude Evans and Robert Stuffelbeam (Dordrecht: Kluwer Academic Publishers, 1997), 141-71.

also "sees" how they would sound differently if they fell to the floor, a clatter not being a thump. Nevertheless, he usually disregards non-visual modalities.

In the ninth place, what Gurwitsch writes about how things are posited and given is probably meager and scattered because he was primarily concerned with what is left when they are abstracted from. The analyses above focused on two species of what are called "noematic characters," "mode of presentation" or "manner of givenness," on the one hand, and "thetic or positional character," on the other. Such characters have been dwelt upon here to clarify just what the difference is between the noematic characters of the concrete thing-as-encountered and the visual noematic core. Once again, there is an identity in contrast with diversity here, this time between core and characters rather than between noema and noeses or between thing encountered and thing-as-encountered. Gurwitsch approvingly refers to "Husserl's distinction between the central noematic nucleus and noematic characters, and … his insistence upon the invariance of the noematic nucleus in the face of variations occurring in the characters" (FC, 363). The nucleus or core of the thing-as-encountered can remain the same while there is change in givenness from being expected to being perceived and to being remembered, and/or when there is change in positedness from being probable to being certain, being liked to being disliked, or from being willed as an end to being willed as a means.

Besides "nucleus," Gurwitsch uses the words "What," "core," and "sense" *(Sinn)* to name what remains of a concrete thing-as-encountered when the noematic characters of givenness and positedness are abstracted from, a point that seems easier to verify if one remains in one place outside the thing and concentrates on visual perceiving and, e.g., the building-as-seen.

A linguistic aspect of the "meaning theory" or "interpretation theory" can now be considered. In English, the Husserlian word *Sinn* is often translated as "meaning," which is naturally comprehended as signifying the signification of a word. That is how Husserl used *Sinn* in the *Logische Untersuchungen* (1900-1901), but in *Ideen I* (1913) and afterwards, he tried to broaden the signification of *Sinn* to include various sorts of sense, and considered *Bedeutung,* which is best rendered as "signification," as one sort. Nevertheless, the older verbal habit often continued even for Husserl, and confusions followed. In English translations, it seems best to render the *Sinn* of the mature Husserl as "sense" and either not to use "meaning" at all or to reserve it to render *Meinung*. Unfortunately, Gurwitsch also regularly uses "meaning" to render *Sinn,* and some colleagues seem to have been misled by this into thinking that his theory of perception includes a linguistic component, which is not the case. Is a translation preference possible regarding *Sinn* for Gurwitsch? There are problems forming an adjective from "What," as there are with the adjective "nuclear," but the third alternative, "core," may be preferred because it can easily be used as an adjective as well as a noun in current American English.

In the tenth place, if what remains of the thing-as-encountered after one abstracts from how it is given and how it is posited is called the "noematic core," how can this core itself be described? Gurwitsch again refers to Husserl: "The expressions to be used for the description of the noematic sense or noematic nucleus may be formal-ontological in character, like 'object,' 'property,' 'state of affairs'; or material-ontological, like 'thing,' 'figure,' 'cause.' They may, furthermore, have material content *(sachhaltig)* like 'rough,' 'hard,' 'colored'" (PTS, 248; cf. SPP, 183).

Because he was chiefly a theorist of logic, Husserl had an emphasis on formal ontological terms, and, within it, on formalization as productive of formally eidetic knowledge. As Gurwitsch puts it, "This procedure consists in replacing all terms of substantive, qualitative, and material denotation by terms, like algebraic symbols, entirely indeterminate as to material content and defined solely by certain relations" (FC, 147; cf. FC, 331). He appreciates his master's emphasis, but seems more concerned with the other type of universalization, i.e., "generalization" (PTS, 267 n. 34), by which one can ascend from this particular green to Greenness, then to the species Color, and then to the genus Sensuous Property (in contrast to other genera of non-sensuous properties, e.g., Intentiveness). The noematic core in the case of looking at a building prominently includes not only "rough" and "hard" as referred to in possible future touching, but the shapes- and colors-as-seen from which there are references to touching.

This result of ten steps of analysis may seem anticlimactic to a reader unfamiliar with Gurwitsch's accomplishment. What we have reached, however, is the point where he can introduce descriptive findings of Gestalt theory into descriptions of noematic cores and thus revise Husserl in fundamental respects. This is not the occasion to summarize all that he takes over from the Gestaltists, which includes extensive interpretations of experimental results and the clarification of such expressions as "Gestalt-character" and "Gestalt-coherence," but something must still be said in relation to looking at a building.

To begin with, and as was evident above in the section on explanation, with the constancy hypothesis abandoned, the thing-as-encountered visually is recognized to be integral. It is a Gestalt, also called a configuration:

> It is an ensemble of items which mutually support and determine one another. Thus they realize a total structure which governs them and assigns to each of them (as a part of the whole) a function or a role to be performed as well as a determinate place in that whole. Each detail exists only at the place at which it plays the role assigned to it by the whole of which it is a part. (SPP, 25)

The side of the building as seen from a fixed outside standpoint already fits this. Within what thus appears, perhaps there is a roof, a wall, and within a wall a door and several windows. The roof, however, would not appear as it does without a wall beneath it, and vice versa; similarly, the door and windows would not appear as they do without each other and the wall without the roof above them. While "roof," "door," and "window" are indeed words with practical connotations, at this point the noematic functional characteristics are abstracted from, so that, although the familiar words are used for convenience, they actually refer only colors and shapes here.

Recalling the discussion in Section I above, it can now be emphasized how the visual appearance is a subtly different appearance when one feature, say the left-most window, is focused on rather than the whole, because of a change in attitude, while the external conditions of stimulation remain the same. On the other hand, should a cloud pass in front of the sun while one is looking, then the external conditions change and so does the core of the noema of looking. Or if there is a row of windows, then the left-most is a "terminal" in relation to the others, which have "intervals" between them, and the windows in a row might (or might not) appear with similarity and equidistantiality (FC, 106 ff.). And in the first passage quoted in the present section, the building stands out from the rest of the street as a figure on a ground (FC, 110 ff.). (More could, of course, be said about colors and their relations with one another and how they are bounded by edges and lines within the configuration.)

What is basic to Gurwitsch's reform of Husserl, then, is that there are no longer hyletic data on which sense is intentively bestowed; instead, there is an integral but structured noematic core. And there is more than what is immediately given in visual perceiving, including the inside of the building as well as its other sides. Gurwitsch describes how they all fit into the configuration that is the building:

> Suppose we perceive a building presenting itself from the front side. That perceptual encounter could not be what it is were it not for references to other sides of the building, not seen at the moment, which contribute toward determining the noematic sense of the present perception as one-sided adumbrational appearance of the building from its front side. Moreover, the front side appears as an architectural form requiring a specific total architectural configuration into which it fits. On the one hand, the perceptual appearance in question arouses, or rather motivates, certain expectancies concerning the total architectural configuration—or, noematically expressed, it predelineates and pretraces the total architectural configuration along more or less indeterminate but specifically generic and typical lines. On the other hand, the total configuration de-

> termines the front side of the building as an architectural detail occupying a specific place within the total architectural configuration. In this sense, the latter may be said to be present in each and every detail; conversely, each and every detail makes its specific contribution toward the architectural configuration as a whole. (PTS, 253)

This is a different account of the relation between the thing-as-encountered and the thing that is encountered than is found in Husserl.

And Gurwitsch clearly goes beyond Husserl when he shows the total field of things-as-encountered always to be structured into a *focus or theme;* a *thematic field or context* of items relevant to the theme; and a *margin* with the important ontological constants of the subject's body, perceptual surroundings, and inner time that have, while marginal, no relevancy to the theme. This fundamental Gurwitschean contention recurs again and again in the chapters of this volume, which now deserve some comments of their own.

IV. Aspects of Gurwitsch Relevant for Cognitive Science

In effect, the contributors to this volume sought aspects of the work of Aron Gurwitsch that were relevant to their own ongoing research. One chapter is pure scholarship (rather than investigation of matters), i.e., my own: I had not expected to contribute more than the introduction to this volume, but it became obvious at the symposium that the central category of relevancy was not as clear, even for me, as I had thought.

The others pursued critical examinations of aspects of Gurwitsch that were relevant to their own investigations, sometimes finding his thought inadequate but always gaining insight from it. If two colleagues had proposed similar topics, I would have urged reconsideration, but, remarkably, this only came close to happening for two chapters, and their convergence only emerged later.

Introductions that merely summarize chapters in the same volume can often be tedious, but a few comments seem worthwhile on what was actually done, including remarks about things I have come to wonder about as well as things I have learned. I defer mention of five additional aspects on which more work might be done to the concluding section.

(1) The first essay is by *Shaun Gallagher* and has been placed there because it is as clear and well written as any other, is by a phenomenologist well informed about cognitive science, and relates to a heretofore under-appreciated aspect of Gurwitsch's work. This aspect is his reflections on encountering other humans, whom he considers originally encountered within practical situations, in their functions, which is to say in their social roles, just as instruments are

practically and situationally encountered in their instrumentality. Gallagher's basis is Gurwitsch's *Habilitationsschrift* of 1931,[10] but the topic can also be found in 1953: "Fellowmen encountered within the sphere of professional activity appear in, and appear as defined by, their roles within that sphere" (FC, 383). Beyond this advance in the phenomenological account of intersubjectivity, however, Gallagher is also concerned with what he calls "primary intersubjectivity."

(2) While Gallagher holds that Gurwitsch advanced the account but still did not reach a completely adequate treatment of intersubjectivity, *Natalie Depraz*, another phenomenological philosopher, finds Gurwitsch's insight quite penetrating where social encountering is concerned, and hints at ethical implications. Her focus is on what Gurwitsch calls "fusion."

In editing these two chapters it occurred to me for the first time that Gurwitsch's account of "signals" in the second essay of *Marginal Consciousness*[11] could be used to describe how the bodies of others, as well as the things they deal with in their actions, signal or indicate psyches in a subcategorial or, in a broad signification, perceptual way. This indicative experience would originate in a subject's consciousness of how the processes in her own psyche indicate processes in her own soma and vice versa. Thus others come to be encountered in indirect consciousness analogous to the way the depicted is indirectly encountered in picture consciousness *(Bildbewusstsein),* but without there being any similarity between bodily movements and mental processes. Again, there is no "meaning theory" or "interpretation theory" in this indirect consciousness as there is in linguistic consciousness. Perhaps someone more specialized than I in "alterology" can see whether this account of signals advances the problems.

(3) *Louis Sass* is a psychopathologist who finds Gurwitsch's category of thematic field illuminating for understanding schizophrenia and other types of mental illness. I am not competent to say more than that this seems plausible to me as a matter of noematic description and that it brings to mind how, as seen in Section I above, past experience affects the brain and then combines with attitudes to explain what is encountered. Post-conference discussion via e-mail with Sass has led me to appreciate more about the category of attitude, something that will be returned to below.

(4) *John Barresi* is a social psychologist who finds great affinity between his own position and that of Gurwitsch with respect to intentiveness and also the ego. Enough has already been said above about intentiveness as a correlation of noesis and noema after one abandons both the constancy hypothesis and the unverifiable distinction between alleged strata within sensuous perceiving (and correlatively within things-as-perceived, e.g., sense data and bestowed "meaning"). About the ego or I, however, it must be said that Gurwitsch also rejects the need, à la neo-Kantianism, for a transcendental ego to organize sense data.

---

[10] See HE, whose original title is *Die mitmenschlichen Begegnungen in der Milieuwelt.*
[11] "The Phenomenology of Signals and Significations," MC, 83-106.

Here Gurwitsch certainly had an unacknowledged indirect influence through Merleau-Ponty on Sartre's "non-egological conception of consciousness" (cf. SPP, Ch. 11). Nevertheless, there is for Barresi an I—or, perhaps better, subject—who is constituted in encountering one's own somatopsychic self and who is in the world, as are others. I believe Gurwitsch would have especially appreciated this convergence of positions.

(5) *Robert Pilat* is a phenomenological philosopher of cognitive science. His topic is the now in inner time, which is presented through Gurwitsch's interpretation of James's description of the stream of thought, in relation to which Pilat shows connections with materials in Husserl's *Nachlass*. The present moment in the stream is intentive (specifically, "protentive") to yet-to-occur mental processes, and also intentive (specifically, "retrotentive") to already-occurred mental processes. The impressional "now" is crucial in this situation. Like the other colleagues in the project, Pilat brings in many results of contemporary cognitive science, including some about neurological conditions.

The recognition that the temporality of the stream of consciousness is not the same as that of the surrounding world in which, e.g., trees grow, is extremely important in the phenomenological tradition. In the natural attitude, the two temporalities are identified, while in the transcendental attitude they are distinguished so that the world can be grounded in a non-worldly consciousness. Such philosophical concerns aside, cognitive science thus has the question of the type of temporality immediately correlated with neurological activity, something that particularly interested Francisco Varela.

(6) Nobody in this project has explored the neurological basis of what Gurwitsch describes more than the young philosopher *Jeff Yoshimi*. He draws on work in analytic-philosophical thought concerning supervenience in order to see how Gurwitsch's field theory of consciousness, which he concisely reconstructs, might be correlated in one or another way with work begun by Wolfgang Köhler on the magnetic fields in the brain. Although I recognize the clarity of Yoshimi's distinctive exposition, I would go beyond even the shallowest edges of my knowledge to say much more about the promise of the hypotheses here for cognitive scientific research. But I am sure Gurwitsch would have been very interested in the experimental results that Yoshimi's analyses could motivate.

(7) *Frédérique de Vignemont* is another young analytic philosopher. Her concern is with the body. Although there are passing remarks elsewhere, Gurwitsch's chief exposition on this topic is in *Marginal Consciousness*. (It may amuse colleagues outside phenomenology to know that there was a time when Husserlians were accused by existential phenomenologists of positing not only solipsistic, but also disembodied and de-situated transcendental egos, which amounts to something of an angelology! Now most phenomenologists know better, and some have even noticed the references to *Ideen II* in Merleau-Ponty.)

It is clear that de Vignemont fully appreciates Gurwitsch's descriptions of the body. This is not the organism constructed in biological science but rather what is sometimes called the lived body.[12] In his ontology, he contends that this body as well as the stream of consciousness and the surrounding world are constants always already there at least in the margin of the field of consciousness (when one thematizes an ideal object, e.g., in mathematical thinking, all three are marginal). But where the organism is concerned, a great deal has been learned since his time. Accordingly, de Vignemont draws on much cognitive science in exploring whether bodily states need to be "objects" even of marginal consciousness, or could instead sometimes be unconscious. She recognizes that the body can be thematic, relevant, and marginal, but also sees a need for the unconscious. More than any other contributor, she seems to have approached her topic from cognitive science. Nevertheless, she interprets Gurwitsch insightfully, appreciating that there is also structure within the margin of the field of consciousness for him, and also that he well recognizes much vagueness and indeterminacy there.

As for the psychological unconscious, I can report that Gurwitsch traced it back to Fechner, whose Gymnasium textbook Freud certainly read, and much appreciated James's objections to it in the chapter of *The Principles of Psychology* on "The Mind-Stuff Theory," chuckling at the statement that the unconscious is the "tumbling ground of whimsy," and agreed with James's suggestion that what is usually called unconscious within the mind is better understood as neurological processes in the brain. Hence the psychology of unconscious processes might be differently interpreted in neuroscience. But this assumes a metaphysical difference between psychological and neurological data. If unconscious and conscious representations (to use cognitive-scientific terms) are both neurological, as a physicalist would hold, then the alternative of James and Gurwitsch does not come to much, something that will be returned to below.

(8) The philosopher *Sven Arvidson* has already written the most thus far on Gurwitsch in relation to cognitive science. At the symposium and in our editorial interactions since then, I was initially more resistant to his position than to that of any other colleague. Eventually, however, I was brought to recognize that our difference was chiefly linguistic. Of course I would not go so far as to say that intentionality is attentionality (or intentiveness is attentiveness), although I fully recognize that the intentive processes said to be attentional (I prefer Gurwitsch's "thematization") belong to a species of intentiveness in general. Nevertheless, I now take the point that if the types of consciousness intentive to the theme, the thematic field, and the margin are described as three species of attention, communication with cognitive scientists could be easier.

---

[12] See Elizabeth A. Behnke, "Edmund Husserl's Contribution to the Phenomenology of the Body in *Ideas II,*" in *Issues in Husserl's "*Ideas II,*"* ed. Thomas Nenon and Lester Embree (Dordrecht: Kluwer Academic Publishers, 1996), 135-60.

(9) *David Woodruff Smith* is an analytic philosopher who has taken Gurwitsch's thought very seriously in an effort to advance his own developments of formal ontology beginning from Husserl. Strictly speaking, the theme/thematic field/margin structure is indeed structural, but—like shape and pattern—it is not formal in the same way as something arising from formalization. Again, Smith is not a Husserl philologist, but a creative philosopher whose leapfrogging over now Gurwitsch as Gurwitsch earlier leapfrogged over Husserl is in many respects fascinating. He has explained to me that he considers Husserl's noematic *Sinn* a technical term referring to intentional content and in many cases expressible in language and thus *Bedeutung*.

(10) Another philosopher, *NOE, Shinya*, relates Gurwitsch's position to other tendencies in the 20th century that emphasize the category of system. I am again too ignorant—this time concerning systems theory—to say much about this essay beyond how intelligently it is written, but I can attest that Gurwitsch was deeply interested in structural linguistics ever since he was, according to Elmar Holenstein, the first phenomenologist to write about it in French,[13] once telling one of my fellow graduate students who had gone to France in the 1960s that he (Gurwitsch) might have become a structuralist if he had not already become a phenomenologist. I am impressed by NOE's interpretations of Husserl as well as of Gurwitsch. Most interesting for me is his discussion of Francisco Varela's work, especially of how transcendental consciousness can be considered an autopoietic system.

(11) I have placed my own essay in the penultimate position because it is merely scholarship and contains no investigation as the other eleven essays do, and will say no more about it here.

(12) The concluding essay is by Michael Schwartz, a psychiatrist, and Osborne Wiggins, another phenomenological philosopher, it concisely presents Gurwitsch's overall position—the big picture one might say—in the theory of science, and I for one find it easy to relate the contents of the other chapters to it. My only difference with them is that I would prefer to say "cultural science" or "cultural discipline" rather than defer to current American academic institutional usage with the expression "social science and humanities."[14] After all and despite many naturalistic tendencies, these disciplines are all concerned with different aspects of the sociocultural lifeworld.

[13] Aron Gurwitsch, "Étude critique de Psychologie du langage," *Revue philosophique de la France et de l'Étranger* 120 (1935): 399-439. Merleau-Ponty is thanked in this article for correcting Gurwitsch's French. Otherwise concerning Gurwitsch's interest in language, see MC, 122 n.10 and Embree, "Introduction" to EPC, 32 ff.

[14] Cf. Lester Embree, "Introduction: Reflection on the Cultural Disciplines," in *Phenomenology of the Cultural Disciplines,* ed. Mano Daniel and Lester Embree (Dordrecht: Kluwer Academic Publishers, 1994), 1-37.

V. Some other yet-to-be Investigated Relevancies of Gurwitsch for Cognitive Science

At least five more aspects of Gurwitsch's work deserve attention from future researchers. (1) As mentioned, Louis Sass led me to ponder the category of attitude further than before. He responded to my comment that the noetic correlate of the thematic field is sometimes called attitude by Gurwitsch (EPC, 144; SPP, 377-80; FC, 238-39; and SPP 203). If one has a taxonomic bent, one might ask the following question. If there is a general consciousness of which the total field is the noematic correlate and "marginal consciousness" and "thematic consciousness" are then species of consciousness intentive to the margin and to the theme respectively, what should the type of consciousness that is intentive to the thematic field be called? "Attitude" is one answer. Another, which some colleagues have already tended toward in their essays, is "contextual consciousness" or "consciousness of context."

That attitudes have a role as internal conditions in the explanation of changes in how things-as-encountered appear has been shown above in Section 1. Gurwitsch does not emphasize the noetic, and hence does not exploit the descriptive potential of the category of attitude as much as Husserl did. This may be because, for Husserl, attitudes pertain to the I, whereas for Gurwitsch, there is no I who performs acts of consciousness. Nevertheless, it would be easy to show that the following classification is implicit in his work.

First, there are practical, aesthetic or valuational, and cognitive or theoretical attitudes. Second, the theoretical attitudes are either straightforward or reflective. Third, the reflective attitudes include those of the natural attitude (for example, in psychology) and those of the transcendental attitude of phenomenological philosophy. In addition, there are attitudes specific to the species and particulars of the cultural as well as the naturalistic sciences within the natural or mundane theoretical attitude. And then there is the contrast between empirical—or, better, factual—and eidetic attitudes, a contrast that holds good in all the disciplines mentioned.

Thus far the taxonomy does not include the attitudes of science-based practical disciplines such as education, engineering, nursing, psychiatry, and psychotherapy. Furthermore, it should be recognized that one can be professionally trained in various theoretical as well as practical attitudes, so that they are habitually assumed in appropriate situations. Finally, there are, on Gurwitschean grounds, different thematic fields correlative to each of the many attitudes or contextual consciousnesses stated or implied in such classifications. After all, disciplines are often alternatively called "fields" and the way issues, data, theories, etc. are included and excluded from the scope of a discipline can be understood in terms of Gurwitsch's theory of relevancy. Much can be done with this category concerning the noematically oriented phenomenology of science and the practical professions.

Then there is a paragraph in Gurwitsch that can be related to attitudes as well as to what he says about the somatopsychic self elsewhere, including his study of Sartre's non-egological conception of consciousness, and it might have implications for cognitive science:

> To be sure, to account for the Ego, one must allow for permanent dispositions, such as attitudes of love, admiration, esteem, hatred, etc., which one person adopts with regard to another person, and for the no less permanent qualities of character and temper, such as likings and dislikings, tendencies, interests, gifts, talents, etc. Both dispositions and qualities have it in common that they may be said to exist permanently, although none manifests itself uninterruptedly. When I feel admiration for a certain person or have a certain leaning, this does not mean that thoughts concerning the admired person or mental states related to the leaning in question exist at every moment in my conscious life. Both dispositions and qualities designate psychic constants, i.e., regularities of experience, action, reaction, behavior, etc., rather than mental facts which themselves fall under direct experience. The logical status of these concepts is much the same as that of physical constants such as index of refraction, electrical and thermal conductivity, specific gravity, etc. Both classes of concepts denote systematic unifications of experienced facts rather than these facts themselves; the unity in both cases is causal and not phenomenological. Concepts of both classes express systematizations and causal unifications of experienced facts through certain mental processes, and it is to these processes and procedures of mind that one must look for an ultimate clarification of the concepts in question. In the last analysis, we are led back to the facts given in immediate experience as materials to be unified and systematized. (MC, 15-16)

It seems plausible that the origins of some, at least, of these dispositions and qualities are to be explained in terms of the causal model presented above in Section I. For example, I did not feel admiration for Aron Gurwitsch before I studied with him but then it became permanent in me. Whether this passage contributes to the aitiology of such conditions as schizophrenia and to accounts in developmental psychology is well beyond my competence, and hence for colleagues such as Louis Sass and John Barresi to ponder.

(2) As a rule, I prepare a paper that I can present if somebody drops out at the last minute from a research symposium. In this case it was doubly fortunate that nobody did, because I did not complete my preparation. But I can say something here about about the topic I had originally envisaged in case somebody wishes to take it up before I can get back to it. Earlier, I followed Schutz in pondering whether others make up a fourth constant in at least marginal consciousness.[15] More recently, I have began to wonder whether ideal objects, and especially universal essences or *eidē,* might not also be ontological constants usually in the margin of the field of consciousness.

It is fairly clear in the earlier writings that Gurwitsch at best considered *eidē* marginal when they had been previously thematized. In his last writings, however, he was increasingly interested in how the typical is encountered and how thematization was then required to separate the *eidos* and the instance that exemplifies it. (Analytical philosophers seem to speak of "type" and "token" in this connection.) Prior to such thematization, one experiences, so to speak, "another car," which is neither "automobile in general" nor this unique particular seen moving down the street. The question is twofold: On the one hand, and philologically, what was Gurwitsch's final position concerning whether the *eidē* exemplified were always already marginal, contextual, or something else prior to such thematization; and, on the other hand, and phenomenologically, how do the things themselves stand in this respect?

(3) Where eideation is also concerned, I had expected that somebody would examine and advance Gurwitsch's work on Goldstein and others where the brain-injured and their difficulties with the categorial and concrete attitudes are concerned. This is discussed to some extent in Gallagher's chapter, but much more can be done. Had somebody taken that topic, I was prepared to report not only that Gurwitsch knew "Schneider," whom some call the hero of the novel entitled *Phénoménologie de la perception,* but also that an archivist at Frankfurt had recently told me that Goldstein's work was challenged by younger psychiatric colleagues after the war; that Goldstein recontacted Schneider, who had, with his wife's help, become the mayor of a small town in Germany; and that that he was able to replicate his results from the 1920s. But, again, I suggest that more can be done by going beyond Gurwitsch's reflections on Gelb-Goldstein.[16]

(4) Another aspect of Gurwitsch that remains untouched has to do with other so to speak "non-normal" subjects, i.e., in comparative psychology (SPP, 67). After all, most of his analyses pertain to civilized, adult, and healthy humans. Probably they are also typically middle or upper class, Western, and masculine. I have pondered how he would have reacted to the suggestion that there were

---

[15] Cf. Editor's Introduction to MC, pp. xxxvi.

[16] I can report, however, that an essay entitled "Gurwitsch, Goldstein, Merleau-Ponty. Analyse d'une droit relation" has been prepared by Profa Maria Luz Pintos at Santiago de Compostela and is forthcoming in *Chiasmi.*

differences in how things and other subjects are encountered according to class, ethnicity, and gender, which involve acquired attitudes, and have come to suspect that he would have been resistant at first, but would finally have accepted what research has shown. His friend Schutz recognized such factors already in the 1950s. In any event, there are interesting remarks about so-called primitives (Gurwitsch knew Lucien Lévy-Bruhl and appreciated his work highly), about children as investigated by Koffka and Piaget, and about chimpanzees whose behavior resembles human behavior in the practical attitude; in fact, Gurwitsch seems the first constitutive phenomenologist to publish remarks on non-human animals (SPP, Ch. 5).

(5) Finally, it may well be that, in order maximally to benefit from cooperation, colleagues researching in the multidiscipline of cognitive science avoid or marginalize metaphysical questions. These could have come up in the discussion of Smith's paper, for example, but did not. Afterwards, some individuals did mention to me that they were physicalists or naturalists. The possibility that both conscious and unconscious "representations" are essentially neurological has been mentioned above. Other colleagues, who consider consciousness to be something that is not physical but rather something mental that can have a non-worldly status would disagree, with the consequence that much of what is said in cognitive science is deeply equivocal. This should be of interest to philosophers *per se*, but, again, may be usefully overlooked in the multidiscipline, i.e., left marginal rather than made thematic or relevant.

*

* *

In sum, a sampling of cognitive scientists and phenomenological as well as analytical philosophers has found Aron Gurwitsch's work relevant for the themes they are interested in and thus added it to their scientific fields. But more can still be done. I again thank the participants for their contributions to our project.[17]

[17] I thank Dr. Elizabeth Behnke for corrections of the form of this essay.

Chapter 1

Situational Understanding: A Gurwitschian Critique of Theory of Mind

Shaun Gallagher
University of Central Florida

Abstract

Gurwitsch's account of how we encounter others provides a corrective contrast to contemporary approaches to social cognition. His analysis is based on phenomenology, but also appeals to psychology and studies of pathology. As such, it shows how there can be phenomenological intervention in the cognitive sciences. This chapter relates Gurwitsch's analysis to contemporary discussions of theory of mind, primary intersubjectivity, and recent discoveries in neuroscience.

Introduction

Contemporary discussions of social cognition and interaction are guided by a mentalistic supposition, namely, that we explain or predict the other's actions and behaviors based on our understanding of their beliefs, desires, or mental states. This evokes the old philosophical problem of other minds, i.e., that although, on some accounts, we have direct, first-person access to our own beliefs and desires, we do not have direct access to those of others. In that case, how can we know what they are thinking, and more generally, understand their intentions? On this supposition, even understanding speech-acts depends on our ability to see that speech expresses mental states. This mentalistic supposition continues to dominate philosophy of mind and the cognitive sciences, despite the constant announcements of a thorough-going rejection of Descartes and everything Cartesian. Specifically, it underpins the two predominate approaches to social cognition, or "theory of mind."

The first approach, "theory theory" (TT), suggests that we "mind-read" the other person's mind by taking a theoretical stance. We *infer* the other's beliefs, desires, and intentions based on a theory (a folk psychology) of how people experience, act and behave.[1] A second approach, called simulation theory (ST), argues that instead of a theory, we use our own mind as a model to simulate

[1]Proponents of TT debate about whether the theory is learned via experience (e.g., Gonik and Meltzoff, 1998), or based on innate capacities that emerge at specific times during development (e.g., Baron-Cohen, 1995).

L. Embree (ed.), Gurwitsch's Relevancy for Cognitive Science, 25-44.

what must be going on in the other's mind. Given what we can see of the other's overt behavior, we consider what it would be like to be in their situation, and we create corresponding simulations or imaginary enactments of "as if" beliefs, desires, or intentional states in our own minds. We then project those mental states onto the other person as an explanation of their behavior, or as a way to predict their behavior.[2]

Both of these approaches find support in experimental science. Theory theory appeals primarily to false-belief tests developed in developmental psychology, showing that children undergo a dramatic change around the age of four years. Prior to the age of four, a child seems unable to grasp the idea that other people may think differently from her, or that they may have false beliefs about certain facts concerning which she knows the truth. Around age four the normal child comes to understand that people may have beliefs or desires that differ from her own. Theory theorists interpret this to mean that prior to age four, children do not have a theory of mind (TOM), but that TOM capacities develop or come "online" around that age (Baron-Cohen, 1995).

Simulation theory is supported by recent neuroscientific studies of mirror neurons and shared neural representations. Mirror neurons in the pre-motor cortex and Broca's area of the human brain are activated both when the subject engages in specific instrumental actions, and when the subject observes someone else engage in those actions. Also, specific overlapping neural areas, including parts of the frontal cortex, are activated under the following conditions: (1) when I engage in intentional actions; (2) when I observe some other person engage in that action; and (3) when I imagine myself or simulate another person engaged in that action (e.g., Grezes and Decety, 2001). These subpersonal mechanisms are said to prefigure our ability to simulate mental states in others (Gallese and Goldstein, 1998).

Many proponents of TT and ST make a strong claim concerning the scope of applicability for their versions of social cognition. The strong claim is that once formed, theoretical stances or simulation routines are our primary and pervasive means for understanding other persons. Changeux, for example, in a recent conversation with Ricoeur, proposed that intersubjective relations depend on a "cognitive device" that allows for the representation of the other's mental states, and he maintains that it is just this type of TOM mechanism that allows humans to acquire a system of moral values and aesthetic preferences (Changeux and Ricoeur, 2000, pp. 154-57). Baron-Cohen (1995, p. 3) writes: "it is hard for us to

---

[2] There are different versions of ST in which simulation routines are either explicit or consciously enacted (see, e.g., Gordon and Barker, 1994), or implicit, automatic, nonconscious, and subpersonal, based on an automatic neurological processing (e.g., Gallese and Goldman, 1998).

make sense of behavior in any other way than via the mentalistic (or 'intentional') framework."[3]

There are good phenomenological objections to be made to both TT and ST, and especially to the strong claim that either or both of these stances constitute our primary and pervasive means for understanding other persons (see Gallagher, 2001). To be clear, the alternative position is not that we never take a theoretical stance towards others, or that we never simulate the mental states of others. Rather, the phenomenological analysis suggests that these stances are relatively rare and specialized in human relations. It is clear, however, that phenomenology by itself, without some account of the experimental science, will not be sufficiently convincing in this regard. To develop a convincing critique or to establish an alternative account, phenomenologists will need to bring scientific evidence to bear on these issues. I want to suggest that the work of Aron Gurwitsch not only offers some resources for this endeavor, but helps to define the right critical path, and in the process provides a model for a phenomenological intervention in the ongoing discussion of social cognition in the cognitive sciences.

I. Gurwitsch's Critique

Aron Gurwitsch (1931) offers a critique of the *argument of inference from analogy*. According to this argument, which can be understood as a forerunner of simulation theory (Gordon and Cruz, 2003), we observe the physical appearance and behavior (including the expressive movements and gestures, etc.) of the other person and then infer their mental experiences, based on an analogy with what we know of our own mental experiences and behaviors. In working out his phenomenological critique, Gurwitsch borrows from developmental psychology and from scientific studies of neuropathology.[4] He engages with early

[3] There are many other examples. Tooby and Cosmides (1995), suggest that 'humans everywhere interpret the behavior of others in … mentalistic terms because we all come equipped with a "theory of mind" module (ToMM) that is compelled to interpret others this way, with mentalistic terms as its natural language' (p. xvii). Also see Wellman (1993); Frith and Frith (1999); and Leslie (2000). Goldman, defending ST, seemingly relaxes this kind of claim. He writes: "Notice that ST need not claim that *all* mentalizing takes the form of simulation, it need only claim that mentalizing often takes that form, or that simulation lies at the roots of mentalizing" (2001, p. 218). But Goldman explains in a footnote (p. 223n11) that this claim is a partial compromise to TT. In other words, whatever ST does not cover, some version of TT does.

[4] Gurwitsch does not put phenomenology into opposition with science. Rather, with respect to psychology he describes part of his own project as follows: to endeavor "to interpret phenomenologically certain psychological theories as well as to use them for the advancement of phenomenological problems. Here, psychology is not looked at from

analytic philosophers, sometimes in agreement—e.g., with Russell's way of distinguishing the psychological and epistemological problems—and sometimes in disagreement—e.g., with Carnap's reduction of expressive movements to the status of purely physical events (Gurwitsch, 1931, pp. 5, 29-30). And although he borrows heavily from phenomenologists such as Max Scheler, Edith Stein, and Martin Heidegger, he does not remain uncritical of them.

Gurwitsch follows and expands on a critique of the *argument of inference from analogy* first offered by Scheler (1954) in 1912. I will outline Scheler's criticisms in five points, and at the same time offer certain qualifications in light of contemporary research.

> 1. In the argument from analogy, the idea that we make inferences to the other person's intentional states is overly cognitive. Scheler suggests that infants and animals, without sophisticated cognition, may be capable of understanding intentions.

There is now good scientific evidence to suggest that infants are indeed capable of understanding the intentions of others. Meltzoff (1995) shows that children at 18 months of age are capable of recognizing and completing another person's failed intention. The experimenter pretends to have great difficulty accomplishing a certain task with a toy, and presents an incomplete action to the child. The child, who takes the toy and completes the task with little effort, demonstrates that she understands what the experimenter desired to do. Whether some animals are capable of understanding intentions is still a debated issue.[5]

> 2. In the argument from analogy, the analogy depends on our ability to compare our own body and bodily behaviors with those of others. Scheler notes that the way my body is experienced by me (e.g., more through kinesthetic-proprioception than through exteroceptive perception) is quite different from the way the body of another person is perceived by me (exclusively through exteroception).

This objection is now questionable. Developmental studies have shown that intermodal perception is operative from the very beginning of life (e.g., Meltzoff and Moore, 1997). What I see is automatically registered in a code that is common to other sense modalities, including proprioception. So when I see the other's body moving in a certain way, I have a kinesthetic-proprioceptive sense

---

without; rather, the psychological theories are received into the philosophical context and are made instruments, not objects, of analysis" (Gurwitsch, 1966, xxi).

[5] In the study that launched the discussion of theory of mind, Premack and Woodruff (1978) argued that chimpanzees were capable of reasoning about the intentions of others. Povinelli (2001) summarizes the evidence against this.

of what that is like. Actually, Husserl (1973) had worked this out in his lectures of 1907. Moreover, it is fully supported by the recent neuroscience of mirror neurons and shared representations mentioned above.[6]

> 3. Scheler suggests that analogy cannot account for our capacity to empathize with animals that have very different kinds of bodies.

This may be a matter of degree, although there is a long tradition in philosophy that affirms the possibility of such empathy.[7] Povinelli (2001) reviews evidence that suggests there is a biological basis for our ability to empathize with other animals. In this regard, Scheler is right to suggest that something less than a cognitive analogy suffices.

> 4. Scheler claims that the argument from analogy is invalid. It unjustifiably opens up an entirely new realm: the *other* mind—observing similar behavior in the other body should only lead us to posit our own mind again.

Theodore Lipps raises the same objection, and Gurwitsch views it as a decisive blow to the theory of inference from analogy (1931, pp. 19ff).[8] This objection strikes me as less than decisive, however. A formal invalidity would exist only if inference from analogy claimed a direct access to the other's mind. Since we are dealing with analogy, rather than a claim of direct access, however, I see no reason why one terminus of that analogy could not be the other mind. The element of alterity is introduced by the other's body, and the element of the mental is introduced by my own mind. My bodily behavior is to my mind as the other's bodily behavior is to *the other's* mind. The logical structure of the analogy seems very much intact.

The intent of this objection highlights an important point. The counter-argument against the argument from analogy should not be that this kind of understanding by analogy is impossible. There are in fact some circumstances where we might rely on just such cognitive inferences—when, for example, it is not at all clear what the strange behavior of the other person really means.

---

[6] For the similarity of Husserl's analysis to the neuroscience of mirror neurons, see Petit (1999).

[7] This tradition includes Darwin, and goes back to Hume who considered this possibility an "evident truth" that "only the most stupid and ignorant" of individuals would deny (cited in Povinelli, 2001, p. 234).

[8] Zahavi (2001) glosses Scheler's objection, without comment, as follows: "Even if all of these [other] problems could be overcome, the argument from analogy would still be formally invalid. Noticing the connection between my own mind and my bodily behaviour, and the analogy between my own bodily behaviour and the behaviour of a foreign body, all that I am entitled to infer is that the foreign body is probably also linked with *my own* body" (p. 152).

Rather, the counter-argument should be that inference from analogy is not the primary or usual way that we come to understand others. To show that this is the case would be sufficient to defeat any strong claim, namely, that inferring from analogy is our common and ordinary way of understanding others.

> 5. Scheler points out that the argument from analogy assumes there is no direct access to the other person's intentions, and this is not the case.

Specifically, Scheler argues that we do have direct access to the others's intentions because those intentions are expressed in their bodily comportment. Scheler appeals to the "phenomenological facts" to justify such access as a direct perception of the other's mind. We get to the other person's mind, not by mind-reading, but by body-reading, that is, by interpreting their actions, gestures, and emotional expressions. This is not a matter of inference, but a matter of perception.

This point, in conjunction with objection (1), turns out to be the point of departure for Gurwitsch's critical analysis. Indeed, his analysis puts this criticism to use in a way that could easily target TT. He suggests, in agreement with Russell, that we have a basic "conviction" that we live in the world with others: "no theoretical or other sort of consideration is even required to arrive at or confirm this 'conviction'" (1931, pp. 3-5). Following both Scheler and Heidegger, he suggests "our *originary* encounter with other human beings does not place us as cognizing subjects over against an object to be cognized (namely, other people)" (p. 35). Furthermore, insofar as the argument from analogy is an early forerunner of ST, Gurwitsch's analysis can be read as critical of the latter. Thus, he writes:

> We certainly do not encounter an absolutely other person, as it were, a second example of the genus to which I belong and which I interpret by analogy or empathy and only make comprehensible by the 'transfer of my own livingness' and its structure, of my own mental processes and their concatenations such that I conceive it in an entirely determinate way as *alter ego*. (1931, p. 108).

Gurwitsch's analysis, however, goes well beyond Scheler. He sets out and criticizes the logical structure of the traditional problem of other minds. The problem has always been framed in terms of the *lack of access* to other minds. He views Scheler's notion of a direct access to the minds of others as an advance over this traditional beginning point. But Gurwitsch contends that the immediate conviction that we live in a world with other people is prior to and foundational for our ability to directly read their expressive actions. This conviction is not reducible to perceptual access to the bodily comportment (expressive movements, gestures, facial expressions) of others, and our knowledge of others is not totally accounted for in terms of our direct access to that expressive movement. Gurwitsch thus wants to expand the realm of phenomena that deliver knowledge of the other person's intentions and feelings.

To do this, Gurwitsch turns to Heidegger's analysis of equipment, circumspective engagement with the surrounding environment, and the larger action contexts of human existence. I think that he achieves an important insight here, and he emphasizes it in a way that Heidegger himself never did: our understanding of another's expressive movements depends on meaningful and pragmatic contexts. Following this path, he moves away from the Schelerian claim that we have a *direct* access to other minds, a claim that is still caught up in the traditional framework that defines the problem in terms of access. He defends the idea that our primary conviction, that we live with others, is mediated by the surrounding world that we are involved in from the very beginning.

Gurwitsch argues that others are not given primarily as objects that we encounter cognitively. We do not, in our primary encounters, begin by *judging* or even thematizing the other in a cognitive act. Rather, our primary encounters are framed in pragmatic contexts. "Prior to all specific cognition, and independent of it, we are concerned with other people in our 'natural living' of daily life; we encounter them in the world in which our daily life occurs" (1931, p. 35). The nature of our *concern* with others, as this word signifies, involves mixing with others in relations that bear on practical matters and interests. It follows that there is not one uniform way in which we relate to others, but that our relations are mediated through various pragmatic circumstances. Furthermore, since we are caught up in such pragmatic circumstances from the very beginning (consider for example the infant's dependency on others for nourishment), the original situation cannot be characterized as an isolated subject confronting an alien being.

> The originary encounter with fellow human beings does not signify a coming together and a being together of isolated individuals who, in their mutual encounter, have severed their collective relations to the surrounding world and, so to speak, find themselves together but detached and "horizonless," as *mere individuals* such that this sort of encounter would be a *mere being together*. Instead, we continuously encounter fellow human beings in a determined horizon. ... (1931, p. 36).

In this determined horizon, we do not have to construct cognitive bridges between our own mind and the mind of the other. As our thinking takes shape, we are already existing in reference to others. The horizon of this reference is such that we do not need to add some social-cognitive relation to the other person on top of our pragmatic concerns in order to make them social, or in order to constitute a separate understanding of the other. "We do not meet other people *alongside of* but rather directly *in* the concrete sector in which we always stand. ... our comportment toward the other is codetermined by our entire situational comportment" (1931, p. 36).[9] This is the alternative approach proposed by Gurwitsch. The primary and originary encounter that we have with others is not only embodied, but embedded, and pragmatic. We meet up with others, either explicitly or implicitly, in the pragmatic contexts of life.

This is true, not only of those others who are closest to us in our pragmatic horizons, but also for strangers, since we encounter them as strangers only in relation to such horizons. This insight leads Gurwitsch to an important principle. "No encounters of fellow human beings are ever given that would be *absolutely* without a horizon such that two 'monads' unrelated to the surrounding world would simply confront each other ..." (p. 36). All such encounters are embedded in situations that already imply an understanding of the other. He offers an important qualification of this principle. It is possible that an encounter with another can transcend this shared horizon. "In that case, being-together finds its meaning in itself, for example, when we are together with a person 'for his own sake'" (p. 36). Accordingly, the horizon is redefined as a purely intersubjective one, as in the case of friendship or love. Still, he cautions, this kind of intersubjectivity, purely for the sake of the other, should not be taken as paradigmatic, and he insists that "it cannot even obtain as the originary and primary sense of being together" (p. 37). The originary and primary sense of being together is always mediated by a pragmatic or concernful living in the surrounding world.

It is this point concerning the intermediary role of the surrounding world that transforms the logic of Gurwitsch's analysis. He suggests that to provide a full analysis of how we understand others, one first needs to provide clarification of this notion of surrounding world. I take this to be a turning point in Gurwitsch's analysis that is both positive and negative. Its positive significance is that it leads him further into his phenomenological analysis of intersubjectivity and the discovery of several important insights. Its negative significance, however, is that it precludes any appeal to intersubjectivity as a primary element in an account of the meaning-consititution of the surrounding world.

Gurwitsch's analysis provides a number of insights. It steers a course between a purely physical and a purely psychical explanation. First, he emphasizes that the body and movements of the other are not just objects or mechanical movements—dead physical things that we must decipher as expressive of mind

---

[9]This can be seen in Meltzoff's (1995) experiment with 18 month-old infants. The context for the experiment was a play and toy setting, and in this familiar setting the infant could easily discern the intentions of the other person who was manipulating an object for a purpose that was quite meaningful.

(1931, p. 52). In line with both Scheler and Husserl, bodily movements are irreducible to the movement of mere objects—they are animate organisms, biologically physical, but more than that since they are expressive of mind. Mind is co-given in the movements of the other person. Second, Gurwitsch takes pains to avoid an overly *cognitive* account of intersubjectivity—e.g., he rejects both Lipps's and Stein's conceptions of empathy as too cognitive (pp. 20, 57). The solution is not to find special cognitive acts that have special empathic insight—the explanation is not to be found in some magical psychical mechanism. In this regard, the primary basis for intersubjectivity is pragmatic involvement rather than cognitive confrontation.

Gurwitsch distinguishes two different conceptions of the *surrounding world*. The first takes its clue from perceptual experience and Husserl's characterization of the "natural attitude." On this view, we perceive others as objects that have the objective status of other things in the world, albeit objects that can be characterized as living. Gurwitsch nicely shows that on this conception of the world the problem of intersubjectivity is precisely the traditional problem of how to gain access to such objects in a way that understands them as human subjects with mental experiences. That is, it is precisely this conception of the surrounding world that motivates the traditional problem, as well as the solutions to that problem that involve arguments of inference from analogy, or approaches like theory theory or simulation theory. This "natural world," however, turns out not to be the world that we actually live in. The world that we actually live in is best captured in the phenomenology of "being-in-the-world" that Heidegger (1927) provides.[10]

> The "being encountered first of all," under which all that is comprehended with which we deal in our daily living, of which we make use, with which we go about our business and manipulate, is immediately delimited by Heidegger in a radical way from physical things as physical things 'on hand' [present-at-hand] that belong to Nature with the material properties and determinations accruing to them. (Gurwitsch, 1931, p. 66).

In contrast to the world of perceived physical things, we are immersed in a practical world that has instrumental significance. To describe this lived world, Gurwitsch refers us primarily to Heidegger's account of the "present-to-hand"

---

[10]Lester Embree has suggested (private communication) that this distinction may have come from Husserl's lecture course on "Natur und Geist" attended by Gurwitsch in 1922 or 1923. Gurwitsch was also in contact with Husserl from 1929 through 1932 and would have heard more about that distinction, given his own interests. Further in the background is the fact that many German philosophers in the 1920's were familiar with similar distinctions made by Dilthey.

nature of equipment (*Zuhandenheit*). But he also supports that account with references to developmental psychology and studies of pathology. Specifically, the development of the naming function in the child's acquisition of language (as explicated by C. and W. Stern [1926]) reflects the fact that children are acquainted not with things in their present-at-hand objectivity, but with how one can interact with them and what one can do with them. Similar evidence can be found in cases of aphasia, where the loss of the ability to use proper names for things may lead to the use of functional descriptions to signify (e.g., instead of "pen" the patient says "for writing") (Gurwitsch, 1931, pp. 68-70).[11]

It is through such scenes of instrumental dealings with the surrounding world that we discover an implicit contact with others. "It is in this milieu that we also encounter our fellow human beings. We encounter them *because* we live in determinate situations of the surrounding world of utensils" (p. 95). Specifically, Gurwitsch suggests that other people are always implicitly on the horizon of our practical dealings with the world. It is the constant and implicit references to others in our pragmatic affairs that generate the inescapable conviction that we live in a human world. This network of implicit references to others is the basis for our experience of being-with, or being together with others. "In these horizonal situations the 'co-included' others appear. That they *come to light in this situation*, and are not 'nearby' or 'merely beside' it, signifies that they appear as belonging to the situation in their specific roles and functions" (p. 97). Here Gurwitsch suggests that our understanding of others is from the beginning in terms of the roles that they play in relation to our projects. "But it is always a matter of a person *in his role*. Understanding is yielded here by virtue of the situation and is, therefore, limited to what is inherent in it" (p. 114).

From the perspective of this analysis, Gurwitsch suggests that our understanding of others is defined in the first place by their roles in pragmatic contexts, and not by their personal individuality or lack of individuality. The others as they appear, co-included in the pragmatic horizons, appear in reference to the pragmatic task, that is, in terms of roles that can be defined sociologically. "So even those 'known personally' are encountered in this reference not as individuals of this determined specific constitution peculiar to them, but instead as sellers, as customers—stated universally, as individuals *in their appropriate roles*" (1931, p. 97-98). The upshot of this account is that in our ordinary everyday encounters, we never come upon others as absolutely nondescript others who conceal a hidden realm of beliefs, desires, or intentions that we need to theorize about or try to simulate. How we understand others is implicitly revealed in the

---

[11]There is a great deal of interesting evidence that could be cited in this regard. Gurwitsch goes on to cite studies of brain damaged patients (a point to which we shall return). For example one could consider the infant's interaction with the world by grasping and buccal exploration, and correspondingly, one could consider utilization behavior following frontal lobe damage.

pragmatic circumstances that define their specific roles. We understand the intentions of others primarily in what they do in the practical contexts that define our relations to them.

The practical context operates as an implicit scaffolding of meaning that rises around us, and shapes our individual encounters. Only when this scaffolding collapses do we find ourselves taking a cognitive and abstract attitude toward the other. Thus, in most cases, I do not need a *theory* to understand the other person; her actions appear meaningful in light of the context that links us. Furthermore, I do not need to simulate a set of beliefs or desires since I can see the meaning of the actions (or inactions) of the other in the pragmatic context.[12] Gurwitsch associates the kind of "implicit situational understanding" gained in the pragmatic context with Dilthey's "elementary understanding" which accesses neither the personal individuality of the other, nor the other's "consciousness-processes, i.e., the *cogitations* of another ego" (1931, p. 112).

> As a consequence, this orientation characterizes this understanding not primarily because it involves other people but because it represents a moment in the knowledge about the immediately actual situation-totality. This orientation grips fellow human beings only insofar as it is a component of the situation. In other words, we are dealing with a functional understanding. (Gurwitsch, 1931, p. 112).

II. Extending the Phenomenological Account

Gurwitsch certainly moves the analysis in the right direction, but does his account go far enough? He explains both the basic conviction that we have about being in a world with others, and the primary way that we go about understanding those others. What he does not explain, however, even in his detailed explication of the appearance of the other on the horizon of our pragmatic projects, the supposed basis for our strong conviction, is precisely how, *in the first place*, the other appears as another person rather than simply another piece of equipment.[13] It is also not clear how just such pragmatic encounters are neces-

[12]"Thus we neither have in front of us—as the position that is the starting point of the traditional inquiry believes [sic]—a 'piece of the external world' in which we empathize, etc., and only when we enliven it, or animate it, or the like, by some form or other of the 'transfer' of our own inwardness, do we then have present another human being ..." (Gurwitsch, 1931, p. 111).

[13]Gurwitsch offers an account of the difference between a tool or instrument and a fellow human being, but only in terms that come too late to count as an answer to how "in the first place" we come to recognize the distinction. The other person is represented as having a freedom to be outside of the situation, whereas the tool has no such possibility (see,

sarily the basis for encounters that are not *just* pragmatic. Motivated by insights provided by experimental science we can ask, "What is it that allows us to engage pragmatically with others—and what is it that prevents this kind of interaction in, e.g., cases of autism?" If, following Gurwitsch, the perceptual experience of another person is not equivalent to the perception of a mere object, is it necessarily equivalent to perception for the sake of instrumental usage? If not, is the difference then totally explainable in terms of Gurwitsch's pragmatic account?

Without giving up the advances made in Gurwitsch's analysis, and his emphasis on the primacy of pragmatic contexts, should we say that social contexts are reducible to or even best understood in terms of pragmatic ones? I want to suggest a negative answer. Gurwitsch is quite willing in his analysis of the pragmatic context to consider the contributions of science, and specifically neuropsychology. He considers a case of apraxia due to brain damage and cites Gelb and Goldstein's distinction between concrete and categorial attitudes, showing the primacy of the concrete (pragmatic) over the categorical, thus supporting a Heideggerian emphasis on the primacy of the pragmatic over the cognitive. In many of Goldstein's own analyses, however, one can find an implicit acknowledgement that the distinction between concrete and categorial attitudes does not draw the complete picture of apraxia, or a variety of other pathologies.

Based on studies of numerous patients, including apraxic-like cases and cases of conduction aphasia, there is good reason to make a three-fold distinction between *abstract* (categorial), *pragmatic* (concrete), and *social* contexts, where the social context is not reducible to the pragmatic or abstract (Gallagher and Marcel, 1999). Patients with motor impairments, for example, show significant improvement in various aspects of motor control and fluency in impaired behaviors when such actions are performed as pragmatically meaningful actions, compared with their performance when elicited as abstract or non-contextualized behaviors. More significantly, in almost all of these cases further measurable improvement was found when nominally the same movements were performed in a socially contextualized situation, in which the movements constituted actions with personal and culturally derived signification.

1931, pp. 116-117). This distinction would seem to be something extra or secondary to what we already accomplish in our situational understanding of the other. In other words, to recognize the freedom of the other person (as a set of possibilities outside of the present pragmatic situation) already implies that we have recognized the other person as a person rather than a tool. Inside of the pragmatic situation where we have our primary encounters with others, as Gurwitsch suggests, "the determination of the role [of the other] and the imprinting of its sense by the whole situation is not essentially different from the way in which a utensil maintains the concrete function accruing to it *hic et nunc*" (p. 117).

> For example, a woman who had difficulty in grasping, in lifting, and in motor fluency when asked in an experimental situation to lift a cylinder of the weight and size of a glass of liquid and to move it toward or away from herself, showed clear improvement in her performance when spontaneously drinking during a meal. This same woman was even more proficient, almost normal, in the very same movements when serving cups of tea to guests in her home, although not when clearing up the cups. (Gallagher and Marcel, 1999, p. 277).

These and other observations suggest that performance in socially contextualized situations is not the same as performance is non-social, pragmatic contexts.

This distinction is reinforced by experiments conducted with a deafferented subject, IW, who lacks proprioception and the sense of touch below the neck. These experiments suggest that, in contrast to Gurwitsch (1931, p. 113), we should not regard expressive phenomena as fully determined by pragmatic contexts. IW has profound difficulties in controlling locomotive and instrumental movement (Cole, 1995; Gallagher and Cole, 1995). To reach and pick up a glass to take a drink, IW is required to keep his hands in his visual field, and to think through his movements. Because of his loss of proprioception, tasks that are normally taken up by body schematic processes are transformed for IW into cognitive tasks. Yet, in spite of this difficulty with pragmatic or instrumental movement, IW is remarkably capable of normal gesticulation in the course of communicative practices with others. IW does not have to think about his hand gestures in order for them to take shape in the normal way. And even when his hands are out of sight, he is capable of close to normal gesture (Cole, Gallagher, and McNeill, 2001). This case, and others like it suggest that social and communicative interactions are not reducible to the pragmatic (instrumental) realm (see Gallagher, Cole, and McNeill, 2001).

In light of these observations about motor performance in social contexts, and the survival of normal gesture when instrumental movement is abnormally controlled, I want to suggest that the distinctions between cognitive, pragmatic, and social aspects of experience are important ones to consider in relation to the question of how we understand one another. Even if normally these aspects are fully integrated, how we understand and interact with one another is still not reducible or fully explainable in either purely pragmatic or purely cognitive terms. *Pace* Gurwitsch, social existence is something more than simply encountering others pragmatically, or in pragmatic contexts, just as it is more than taking a theoretical stance towards others.

III. Primary Intersubjectivity

Gurwitsch's critique of the argument of inference from analogy works equally well against theory theory and simulation theory. His own positive proposal advances the importance of pragmatic contexts for understanding how we interrelate and understand each other. I have suggested, however, that he has not gone far enough towards recognizing the uniqueness of the social. He is right, however, that the solution is not to look for special empathic cognitive acts or some magical psychical mechanism. The alternative, I think, is to be found in what Trevarthan (1979) calls *primary intersubjectivity*.

Prior to the age of four, when children begin to recognize that others are capable of having false beliefs, human interaction and intersubjective understanding have already been accomplished in certain embodied practices—practices that are emotional, sensory-motor, perceptual, and nonconceptual. These embodied practices constitute our primary access for understanding others, and they continue to do so even after we attain theory-of-mind abilities (Gallagher, 2001).

In most intersubjective situations we have a direct understanding of another person's intentions because their intentions are explicitly expressed in their embodied actions and their expressive behaviors. This understanding does not require us to postulate a belief or a desire hidden away in the other person's mind. What we reflectively or abstractly call their belief or desire is expressed directly in their actions and behaviors.

A primitive sense of others is already implicit in the behavior of the newborn. In neonate imitation, infants are able to distinguish between inanimate objects and people—they imitate only *human* faces (see Legerstee, 1991; Johnson, 2000; Johnson et al., 1998; Slaughter and McConnell, 2000). An intermodal communication between proprioception and vision is already functioning at birth. There is a common bodily intentionality that is shared across the perceiving subject and the perceived other. From the very beginning, in contexts that involve expressive movement and effortful strivings, but outside of and prior to the infant subject's experience of meaningful pragmatic circumstances, infants are able to parse the surrounding environment into those entities that perform human actions (people) and those that do not (things) (Meltzoff and Brooks, 2001).

Other early interactive capabilities that support a more immediate, non-mentalistic mode of interaction contribute to primary intersubjectivity. The ability to read another's intentions is found early in infancy (Baron-Cohen, 1995, p. 32). Infants, notably without the intervention of theory or simulation, are able to interpret bodily movement as goal-directed intentional movement, and to perceive other persons as agents. This does not require advanced cognitive abilities; rather, as Scholl and Tremoulet suggest, it is 'fast, automatic, irresistible and highly stimulus-driven' (2000, p. 299).[14] This capability may manifest itself in pragmatic contexts, but is not reducible to an understanding that is initiated in instrumental terms (Meltzoff, 1995, 1996, 1999). Just as the neonate imitates only human faces, as early as 15-months of age the infant recognizes and responds to the intentions of other persons, but not to similar movement of a mechanical apparatus. This suggests that the infant does not understand the intentions of the other primarily through the context of instrumentality, but by seeing the other's human body gesture and act. That the activity is goal-directed is not perceived as such primarily through the specific instrumental tasks, but in the specifically human actions that accomplish or fail to accomplish those tasks.

Of course children do not simply observe others, they interact with them in ways that are not simply pragmatic. They perceive and share attention with others.[15] Well before the ability to pass false belief tests at four years, the child looks to the body and the expressive movement of the other to discern the intention of the person or to find the meaning of some object.

This perception-based understanding is a form of *body*-reading rather than *mind*-reading. In seeing the actions and expressive movements of another, one already sees their meaning; no inference to a hidden set of mental states (beliefs, desires, etc.) is necessary. Primary intersubjectivity also includes affective coordination between the gestures and expressions of infants and those of the other persons with whom they interact (Gopnik and Meltzoff, 1997). Infants at five to seven months detect correspondences between visual and auditory information that specify the expression of emotions (Walker, 1982). The understanding of emotion in the movement of others, however, does not involve taking a theoretical stance or creating a simulation of some inner state. It is a perceptual experience of an embodied comportment.[16]

The capabilities involved in primary intersubjectivity suggest that before we are in a position to form a theory about, to analogize, or to simulate what the other person believes or desires, we already have specific perceptual understand-

---

[14]Evidence for this early, non-mentalistic interpretation of the intentional actions of others can be found in numerous studies. Baldwin and colleagues have shown that infants at 10-11 months are able to parse some kinds of continuous action according to intentional boundaries (Baldwin and Baird, 2001; Baird and Baldwin, 2001).

[15]Behavior representative of joint attention begins to develop around 9-14 months (Phillips, Baron-Cohen, and Rutter, 1992).

[16]Moore, Hobson, and Lee (1997) demonstrated the emotional nature of human movement using actors with point-lights attached to various body joints. As early as five months of age infants show preferential attentiveness to human shape and movement in such displays (Bertenthal, Proffitt, and Cutting, 1984). Hobson (1993) cites Merleau-Ponty (1994) who notes the "simple fact that I live in the facial expressions of the other, as I feel him living in mine" (p. 146). Also see Cole (1998, 1999) on the importance of the face in such contexts.

ings about what people intend in particular contexts. Moreover, these particular contexts are not necessarily pragmatic ones. We interact with and understand others emotionally in terms of their gestures and intentions and in terms of what they see, what they do or pretend to do with objects, and how they act toward ourselves and others. Furthermore, primary intersubjectivity is not primary simply in developmental terms. Rather, it remains primary across all face-to-face intersubjective experiences. It subtends what is only the occasional and secondary intersubjective practices of explaining or predicting mental states in others. Accordingly, it undermines the strong claims made by TT and ST, that either a theoretical stance or simulation constitutes the primary and pervasive way we have of understanding others.

Conclusion

Although I have argued that Gurwitsch's emphasis on the pragmatic does not tell the entire story, it is clearly right to say that we often understand the intentions of others precisely within pragmatic contexts. And when we do so, we understand them, not in purely physicalistic terms, nor in purely mentalistic terms. We do not interpret actions, our own or others, on either an abstract, physiological level ("I am [she is] activating a certain group of muscles"), or in terms of a mentalistic performance ("I [she] believe[s] P, so I [she] will do X"). Rather, quite naturally, we understand actions on the highest pragmatic level possible (see, e.g., Jeannerod, 1997; Gallagher and Marcel, 1999). For example, if, as I reach for a glass, someone asks me what I am doing, I do not say, ordinarily, "I am reaching for a glass"; rather I say, "I'm taking a drink." We tend to understand actions just at that pragmatic, intentional (goal-oriented) level, ignoring possible sub-personal or lower-level descriptions, but also ignoring ideational or mentalistic interpretations. When I'm asked, "What are you doing?" I do not respond, "I'm acting on a belief that I am thirsty." Likewise, the interpretation of the actions of others occurs at that same pragmatic (intentional) level. We interpret their actions in terms of their goals and intentions set in contextualized situations, rather than abstractly in terms of either their muscular performance or their beliefs.

To get this kind of pragmatic interpretation "off the ground," however, we already need to be keyed into the significance of others as distinguished from objects. As Gurwitsch rightly contends, this does not depend on a "magical psychical mechanism." Citing evidence from the developmental and cognitive sciences, I have suggested that the ability to tune into the significance of others is part of the innate and early developed capacities which constitutes primary intersubjectivity. Social interaction and the understanding of others should not be reduced to an interaction between objective bodies, or inflated to an interaction between minds. Nor should we give up his account of how our understanding of the other's expressive movements and intentions are accomplished in meaningful pragmatic contexts. Thus, the strong claims of theory theory and simulation theory concerning their scope, can be rejected in favor of the notion of a primary intersubjectivity that is carried forward into precisely the kind of "implicit situational understanding" that Gurwitsch describes.

References

Baird, J. A. and Baldwin, D. A. 2001. Making sense of human behavior: Action parsing and intentional inference. In B. F. Malle, L. J. Moses, and D. A. Baldwin (eds). *Intentions and Intentionality: Foundations of Social Cognition* (pp. 194-206). Cambridge, MA: MIT Press.

———. 2001. Discerning intentions in dynamic human action. *Trends in Cognitive Science* 5 (4): 171-78.

Baron-Cohen, S. 1995. *Mindblindness: An essay on autism and theory of mind.* Cambridge, MA: MIT Press.

Bertenthal, B. I. Proffitt, D. R. and Cutting, J. E. 1984. Infant sensitivity to figural coherence in biomechanical motions, *Journal of Experimental Child Psychology*, 37, 213-30.

Changeux, P. and Ricoeur, P. 2000. *What Makes us Think?* (trans. M. B. DeBevoise). Princeton: Princeton University Press.

Cole, J. 1999. On "being faceless": Selfhood and facial embodiment. In S. Gallagher and J. Shear (eds). *Models of the Self* (pp. 301-318) Exeter: Imprint Academic.

———. 1998. *About Face*, Cambridge, MA: MIT Press.

———. 1995. *Pride and a Daily Marathon.* Cambridge, MA: MIT Press.

Cole, J., Gallagher, S., and McNeill, D. 2002. Gesture following deafferentation: A phenomenologically informed experimental study. *Phenomenology and the Cognitive Sciences*, 1 (1): 49-67.

Decety, J. 2002. Naturaliser l'empathie. *L'Encéphale.* 28: 9-20.

Frith, C. D. and Frith, U. 1999. Interacting minds - A biological basis. *Science,* 286, 1692-1695.

Gallagher, S. 2001. The practice of mind: Theory, simulation, or interaction? *Journal of Consciousness Studies*, 8 (5–7): 83–107.

———. 1996. The Moral Significance of Primitive Self-Consciousness. *Ethics,* 107, 129-140.

———. and J. Cole. 1995. Body schema and body image in a deafferented subject. *Journal of Mind and Behavior* 16: 369-390.

Gallagher, S., Cole, J. and McNeill, D. 2001. The language-thought-hand system. Oralité et gestualité: Interactions et comportements multimodaux dans la communication.. Eds. C. Cave, I. Guaitella, and S. Santi. Paris: L'Harmattan: pp. 420-24.

Gallagher, S. and Marcel, A. J. 1999. The self in contextualized action. *Journal of Consciousness Studies,* 6 (4), 4-30.

Gallagher, S. and Meltzoff, A. N. 1996. The earliest sense of self and others: Merleau-Ponty and recent developmental studies. *Philosophical Psychology,* 9, 213-236.

Gallese, V. 2001. The 'shared manifold' hypothesis: from mirror neurons to empathy. *Journal of Consciousness Studies* 8: 33-50

Gallese, V. and Goldman, A. I. 1998. Mirror neurons and the simulation theory of mind reading, *Trends in Cognitive Science,* 2, 493-501.

Goldman, A. I. 2001. Desire, intention, and the simulation theory. In B. F. Malle, L. J. Moses, and D. A. Baldwin (eds). *Intentions and Intentionality: Foundations of Social Cognition* (pp. 207-224). Cambridge, MA: MIT Press.

Gopnik, A. and Meltzoff, A. N. (1997. *Words, Thoughts, and Theories*, Cambridge, MA: MIT Press.

Gordon, R. M. and Cruz, J. 2003 (in press). Simulation theory. *Encyclopedia of Cognitive Science.* London: Nature Publishing.

Gordon, R. M. and Barker, J. A. 1994. Autism and the theory of mind debate. In G. Graham and G. L. Stephens (eds). *Philosophical Psychopathology* (pp. xxx). Cambridge, MA: MIT Press.

Grezes, J. and Decety, J. 2001. Functional Anatomy of Execution, Mental Simulation, Observation, and Verb Generation of Actions: A Meta-Analysis. *Human Brain Mapping*, 12: 1–19.

Gurwitsch, A. 1931/1978. *Human Encounters in the Social World.* Pittsburgh: Duquesne University Press.

———. 1966. *Studies in Phenomenology and Psychology* (Evanston, IL: Northwestern University Press.

Heidegger, M. 1927. *Sein und Zeit [Being and Time].* Halle.

Hobson, P. 1993. The emotional origins of social understanding. *Philosophical Psychology,* 6: 227-49.

Husserl, E. 1973. *Ding und Raum: Vorlesungen, 1907.* Husserliana 16. Ed. U. Claesges. Hague: M. Nijhoff.

Jeannerod, M. 2001. Neural simulation of action: A unifying mechanism for motor cognition. *Neuroimage*, 14, S103-S109.

———. 1997. *The Cognitive Neuroscience of Action.* Oxford: Blackwell Publishers.

Johnson, S. C. 2000. The recognition of mentalistic agents in infancy. *Trends in Cognitive Science*, 4, 22-28.

Johnson, S. et al. 1998. Whose gaze will infants follow? The elicitation of gaze-following in 12-month-old infants. *Developmental Science*, 1, 233-238.

Legerstee, M. 1991. The role of person and object in eliciting early imitation. *Journal of Experimental Child Psychology* 51: 423-433.

Leslie, A. 2000. Theory of mind as a mechanism of selective attention. In M. Gazzaniga (ed). *The New Cognitive Neurosciences* (pp. 1235-47). Cambridge, MA: MIT Press.

———. 1994. ToMM, ToBy, and Agency: Core architecture and domain specificity. In L. Hirschfeld and S. Gelman (eds). *Mapping the Mind: Domain Specificity in Cognition and Culture.* Cambridge: Cambridge University Press.

Meltzoff, A.N. 1995. Understanding the intentions of others: Re-enactment of intended acts by 18-month-old children. *Developmental Psychology* 31: 838-850.

———. 1996. Understanding intentions in infancy. Symposium paper. 26th International Congress of Psychology, Montreal.

———. 1999. Origins of theory of mind, cognition, and communication. *Journal of Communication Disorders* 32: 251-269.

Meltzoff, A. N. and Brooks, R. 2001. 'Like me' as a building block for understanding other minds: Bodily acts, attention, and intention. In B. F. Malle, L. J. Moses, and D. A. Baldwin (eds). *Intentions and Intentionality: Foundations of Social Cognition* (pp. 171-91). Cambridge, MA: MIT Press.

Meltzoff, A. and Moore, M. K. 1997. Explaining imitation: A theoretical model. *Early Development and Parenting.* 6: 179-92.

Merleau-Ponty, M. 1962. *Phenomenology of Perception*, trans. C. Smith. London: Routledge and Kegan Paul.

———. 1964. *The Primacy of Perception*, translated by W. Cobb. Evanston: Northwestern University Press.

Moore, D. G., Hobson, R. P. and Lee, A. (1997. Components of Person Perception: An investigation with autistic, non-autistic retarded and typically developing children and adolescents. *British Journal of Developmental Psychology,* 15, 401-423.

Petit, Jean-Luc. 1999. Constitution by movement: Husserl in light of recent neurobiological findings. In J. Petitot et al. (eds). *Naturalizing Phenomenology: Issues in Contemporary Phenomenology and Cognitive Science* (pp. 220-44). Stanford: Stanford University Press.

Phillips, W., Baron-Cohen, S. and Rutter, M. 1992. The role of eye-contact in the detection of goals: Evidence from normal toddlers, and children with autism or mental handicap. *Development and Psychopathology*, 4, 375-383.

Povinelli, D. J. 2001. On the possibilities of detecting intentions prior to understanding them. In B. F. Malle, L. J. Moses, and D. A. Baldwin (eds). *Intentions and Intentionality: Foundations of Social Cognition* (pp. 225-48). Cambridge, MA: MIT Press.

Premack, D. and Woodruff, G. 1978. Does the chimpanzee have a theory of mind? *Behavioral and Brain Sciences* 1: 515-26.

Scheler, M. 1954. *The Nature of Sympathy.* Trans. Peter Heath. London: Routledge and K. Paul. (Original: 1912/1973. *Wesen und Formen der Sympathie.* Bern: Francke Verlag).

Scholl, B. J. and Tremoulet, P. D. 2000. Perceptual causality and animacy. *Trends in Cognitive Science,* 4 (8), 299-309.

Slaughter, V. and McConnell, D. 2000. *Differential imitation of human and nonhuman models by older but not younger infants*. Poster presented at the International Conference on Infant Studies, Brighton, England.

Todes, S. 2001. *Body and World.* Cambridge, MA: MIT Press.

Tooby, J. and Cosmides, L. 1995. 'Foreword' to S. Baron-Cohen, *Mindblindness: An essay on autism and theory of mind* (pp. xi-xviii). Cambridge, MA: MIT Press.

Trevarthen, C. 1979. Communication and cooperation in early infancy: A description of primary intersubjectivity. In M. Bullowa (ed). *Before Speech.* Cambridge: Cambridge University Press.

Walker, A. S. 1982. Intermodal perception of expressive behaviors by human infants. *Journal of Experimental Child Psychology*, 33, 514-35.

Wellman, H. M. 1993. Early understanding of mind: The normal case. In S. Baron-Cohen, H. Tager-Flusberg, and D. J. Cohen (eds). *Understanding other minds: Perspectives from autism.* Oxford: Oxford University Press.

Zahavi, D. 2001. Beyond empathy: Phenomenological approaches to intersubjectivity. *Journal of Consciousness Studies* 8 (5-7): 151-67.

Chapter 2

Vertical Context after Gurwitsch

Natalie Depraz
Université de la Sorbonne (Paris IV)

Abstract

To what extent is Aron Gurwitsch's understanding of context inherently intersubjective? We will try to answer this question by appealing to the phenomenological experience of horizonality, then by exploring the light paradigm as it is thematized by different phenomenologists. Inasmuch as the mystical experience leads to reconsidering our access to perception as well as its limitations, we will rely on it to show how Gurwitsch truly paves the way for a genuine open understanding of context that is able to renew in turn the very experience of intersubjectivity: this is what we here call "vertical context."

Introduction

Aron Gurwitsch is well known for his analysis of the *structure* of the field of perception and his understanding of the intentional correlation of consciousness and its total field. In his classic work, *The Field of Consciousness,* he raises this issue impressively by setting forth the notion of *context*:

> To develop a field-theory of consciousness is to embark upon an analysis of the phenomenon of context in general, as well as upon the eventual disclosure of different types of contexts. By this we mean types which are distinguished from one another by virtue of differences involving the organization principles which prevail in the several types. Context will be understood as experienced context, that is, as that which presents itself to the experiencing subject's mind. Therefore our analysis will be descriptive in character. We shall endeavor to account for context in terms of experience of context.[1]

[1] Aron Gurwitsch, *The Field of Consciousness* (Pittsburgh: Duquesne University Press, 1964), p.2. Hereafter this source will be cited textually as "FC."

L. Embree (ed.), *Gurwitsch's Relevancy for Cognitive Science*, 45-57.

In the previous paragraph he defines the field of consciousness as the "totality of co-present data," with co-presence "understood in a broad sense so as to comprise not only data which are experienced as simultaneous but also those which are simultaneously experienced, though not as simultaneous" and characterizes the task of a field theory of consciousness as "the task of investigating the articulation of the total field of consciousness and of bringing out the patterns and forms in which co-present data are organized with respect to each other" (FC, 2). The organization of the field of consciousness was already described in *Marginal Consciousness* through reference to the co-presence of marginal data with the theme and the thematic field as a relation between intentional correlates of concurrent mental acts.[2]

Besides Gurwitsch's understanding of the context as a totality of co-present data related to the very lived act-experience of the subject, he considers that subject embodied: "With all due allowance for the indubitable difference between the psychic and somatic, the Ego may be taken as a unity, as the psycho-somatic unity it is given as in everyday experience" (MC, 26). Mental acts refer to something spatial and temporal as well as to each other in time. There is a kinetic-bodily simultaneity and an immediate successiveness inherent in the phenomenal act of being present, which includes what just happened and what is about to occur.

In the light of such an analysis of consciousness recognizing the spatio-temporal foundation of perceptual acts, our question is the following: to what extent does the Gurwitschean notion of context do justice to the intersubjectivity of experience? Does intersubjective experience derive from the spatio-temporal foundation of perception and is ethics therefore secondary with regard to the primacy of a cognitive perception or is intersubjectivity an intrinsic dimension of perception, which then compels us to differentiate between different forms of intersubjectivity? And to what extent is the meaning of intersubjectivity changed if it is not understood not merely in the framework of a horizonal context but also in the light of what we will call a "vertical context?"

We will proceed in three steps: 1) we will show how Gurwitsch's notion of context provides us with a genuinely *cognitive* concept of intersubjectivity that is linked to but also differs from the Husserlian notion of horizon; 2) we will engage in an analysis of light and luminosity in order to pave the way for a description of context that enables us to emphasize its depth-dimension and reveal its tri-dimensionality, its relief, which will in turn lead us to consider a broad-

---

[2]Aron Gurwitsch, *Marginal Consciousness*, Ed. Lester Embree (Athens, OH: Ohio University Press, 1985), p. xliv: "The co-presence of marginal data with the theme and the thematic field is only a relation between the mental acts, i.e., the experiences having such matters as intentional correlates, and this relation consists solely in these acts happening to occur together." Hereafter this source will be cited textually as "MC."

ened *pragmatic* understanding of intersubjectivity, as it is labeled, for example, by Gurwitsch in the expression of "consociate being together"; and 3) we will finally try to make sense of the possibility of a "vertical context" emerging from its deepened horizontality and exceedingly trespass it from within, which brings about a *radically embodied* concept of intersubjectivity defined by its dynamics of self-transcendence and involving a truly relational ethics.

I. Context and Horizons: A Constitutive/Cognitive Intersubjective Horizontality of the Phenomenon

We need first to make an initial distinction between two main understandings of intersubjectivity: 1) in its narrow understanding (but also its strongest, more accurate, and thus more rigorous meaning), intersubjectivity is equivalent to empathy and 2) in its broad (but also more diffuse, evasive, and vague) meaning, it is synonymous with the capacity to be open to the world and to others. In other words, while the first intersubjectivity is a method that is focal, reflective, and formal, the second one is more basic, spontaneous, and ecological. Being situated at two different levels, these two understandings of intersubjectivity are complementary and both need to be taken into account: 1) empathy is a well-structured and neatly-layered experience that nicely fits concrete and precisely defined empirical protocols and 2) the capacity for openness rather meets the requirement of a more general epistemological methodology, which also enables redefining our perceptive activity.

In that latter respect, it may be useful to refer on the holistic methodology as it is remarkably described by Kurt Goldstein in *Der Aufbau des Organismus*:

> The following is the first methodological requirement: *One must take account of all the phenomena that an organism, e.g., a patient, presents and give preference not from the outset to any of them in the description that one makes.* Initially, there are neither important nor insignificant phenomena. ... A phenomenon can be extremely significant when it receives nearly no attention. ... *The second methodological requirement* is to clarify the phenomena themselves. In this respect, one often errs by confining oneself to simply recording an effect, but an effect is *equivocal* in relation to the function below it. Only a deep analysis of the genesis of a success as much as a failure makes it possible to see clearly. ... *Negative* results are as equivocal as positive results; they too can only be clarified by a deep analysis. Too often one has the habit of recording them as simple blockages, whereas in reality when one submits it to a meticulous analysis, they cast a bright light on the behavior

> of the organism, they are *false responses*. ... Our *third methodological requirement* is to consider any phenomenon without reference to the organism and the situation in which one observes it.[3]

The biologist provides us with a handsome "*précis* of methodology" that can be summarized in three main rules: 1) the experiencing subject needs to be open to all aspects of a given phenomenon: this implies an attitude without presuppositions, that is, an *epochē* of every prejudice. Consequently, one avoids taking one particular aspect into account just because one does not know what to do with it or because it is astonishing or disturbing. The recommended attitude is, on the contrary, to become interested precisely in such an aspect, which is all the more meaningful if it first seems unintelligible or unable to be integrated. Thus one needs to develop a capacity for openness toward the whole of the phenomenon;

2) Such an open attitude goes hand in hand with attention to the multidimensionality of the experienced phenomenon, namely, to its very situatedness or individuation. What is here at stake is precisely the component of the context or horizon4 or of the environment or milieu[5] as an intrinsic dimension of the phenomenon and not as something exterior to it; and

3) Hence the final requirement of proceeding to a completely categorizing description that is attentive to and aware of the various descriptive modalities and structural levels of the phenomenon. This categorizing indication is a direct consequence of the first two more experiential ones and is completely coherent with them. Characterized by the criterion of openness, Goldstein's methodology thus completely fulfills the requirement of a concept of phenomenon inherently constituted by the cognitive intersubjective structure of horizontality.

In his *Studies in Phenomenology and Psychology* and in his *Esquisse de la phénoménologie constitutive*,[6] as well as in *The Field of Consciousness*, espe-

---

[3] Translating from Kurt Goldstein, *La structure de l'organisme* (original German version 1934 under the title: *Der Aufbau des Organismus*), Paris, Gallimard, 1951, p. 20-24. See also V. Rosenthal and Yves-Marie Visetti, « Sens et temps de la Gestalt », *Intellektica*, 1999/1, p. 147-227.

[4] Aron Gurwitsch, *op. cit.*, and Edmund Husserl (1913), *Hua* I, 1950.

[5] Edmund Husserl (1934/1937), *Hua* VI, 1966, and Jacob von Uexküll, *Mondes animaux et monde humain*, Paris, Denoël, 1957.

[6] Aron Gurwitsch, *Studies in Phenomenology and Psychology,* Evanston, Northwestern University Press, 1966, chapter 10: "Phenomenology of Thematics and of the Pure Ego: Studies of the Relation between Gestalt Theory and Phenomenology," pp. 175-286; also, *Esquisse de la phénoménologie constitutive* (Paris: Vrin, 2002). As far as Goldstein is concerned, see *Esquisse*, Appendice V: "La science biologique d'après K. Goldstein," pp. 369-389.

cially the first part, Aron Gurwitsch frequently refers to Gestalt theory and to Kurt Goldstein. While never discussing empathy, i.e., the first Husserlian concept of intersubjectivity, which involves the transcendental attitude and the strong notions of reduction and constitution, Gurwitsch's understanding of consciousness and its field is completely coherent with the broader notion of intersubjectivity understood as openness to the world. In that respect, contextual consciousness may be said to be equivalent to an inherently intersubjective consciousness. Already in *Marginal Consciousness*, he makes use of notions such as "horizonal consciousness": "Horizonal consciousness is never awareness of something utterly and absolutely uncertain and random but is rather an extension along determinate lines, however schematically delineated" (MC, 43). This seems to imply that such a consciousness is to some extent structured even though it is not thematic.

Such a plastic structuring of the contextual *qua* horizonal consciousness does not amount to the looseness of the marginal consciousness, which is, for example, experienced in the kind of awareness we have of our bodily existence: "a certain dim, vague, indistinct, and indeterminate awareness of our 'embodied existence' accompanies us throughout our conscious life." Then again, "With regard to our definitions and distinctions betweens the items which are and those which are not materially relevant and pertinent to the theme (assuming both copresent with the theme), we must classify this concomitant awareness of bodily conditions under the heading of 'marginal consciousness'"(MC, Ch. 3, pp. 27 & 30). In addition, "horizonal awareness" is the very consciousness of such a horizon: "We become aware of the infinite openness of the horizon and of the 'world' as the all-embracing horizon which extends infinitely in the performance of the process of clarification" (MC, 45).

So it seems that consciousness and awareness involve interchangeable experiences: the real difference is between a *marginal* consciousness of a dim and indeterminate awareness founded in the body, thus an essentially inner or intrasubjective awareness, and a *horizonal or contextual* one characterized as openness to the world and thus essentially of a cognitive intersubjective type. But the notion of horizon itself appears to be broader and still less determinate:

> The experiences of bodily facts refer and point, furthermore, beyond these facts to an inarticulate and confused horizon which, when apprehended and unfolded in the appropriate attitude, displays itself in a coherent and systematic group of bodily facts, both actual and virtual, the totality of the latter actually defining the bodily condition of the subject at a given moment. (MC, 34-35)

Thus, insofar as it might be completely inarticulate and thus not internally determined, the Gurwitschean horizon itself is not necessarily intersubjective in the sense of an openness to the world. Ultimately, the experience of contextuality alone provides us with such a genuine cognitive intersubjective openness.

II. The Light Paradigm: Broadening Contextual Cognitive Intersubjectivity through the Pragmatics of Being Together

If context and potentially intersubjective co-presence are narrowly linked, the consciousness of field is ruled by a structure of differentiation and heterogeneity. Thus, as intimated already, I am more or less conscious of what happens to me, I am able to become aware of certain bodily aspects of myself that I did not notice initially, some of them remaining fully inaccessible to me, while others become understandable.

The metaphor of the clarity of consciousness, of its obscure dimensions and of its elucidation or clarification, is commonly used in reference to this heterogeneous complexity in our capacity to be aware or not. This image of light, luminosity, and opacity is also commonly used by psychologists, by phenomenologists, and by cognitivists in order to understand the very structure of perception. It seems to be a transversal metaphor for our inner realm as well as for our perceptive relationship to the world.

To what extent is such an image useful in order to better understand the Gurwitschean notion of context, which provides a key in order to link inner consciousness and the outer structure of perception? How can such a light metaphor be of use in order to deepen the idea of "intersubjective context?"

As a focus of visibility and of vision, the notion of light has been used and thematized by various phenomenologists for at least two reasons: 1) some emphasize the ambiguity of vision in the framework of our perceptive activity or in the practical background of our skills or know-hows that is both an evident source of knowledge and a testimony for our limited ability to full clarity and 2) others stressed the immersion of our sensuous visual activity into tactile invisibility.

While describing these different phenomenological approaches to the visible and invisible and the understanding context as an intersubjective mediation between consciousness and its surrounding world, we hope to show how far they help us to situate Gurwitsch's analysis better.

A) The chiaroscuro of opacity and transparency. Our visual power is relative, on the one hand, to the quality of luminosity of the external environment and, on the other hand, to our inner capacity for vision. The structure and the dynamics of visibility are therefore mainly gradual and mobile and situated halfway between total obscurity and full clarity. In that respect, the latter limit experiences seem abstract because of their irreality and their absoluteness. We are far closer with intermediate and standard modalities of visibility, much more akin to our visual capacity, and to the surrounding possibilities of luminosity.

Such relative qualities of luminosity are first and foremost stressed by the founding fathers of phenomenology, i.e., Husserl and Heidegger: the first considers perceptual activity as a gradual continuum of shadow and light, while the second is more interested in the various forms of the transparency of our know-hows.

The structure of perception is conceptualized by Husserl thanks to the notions of "profiles or adumbrations (*Abschattungen*)" in their very givenness and of "inner and outer horizons." The image of the glance (*Blick*) plays the part of orienting and channeling vision while focusing on a particular object in the given field. Such a focal object of is thus brought to full light, properly illuminated, whereas the surrounding remains in shadow, i.e., co-perceived or co-given without being directly looked at.[7] This preliminary main distinction is directly taken over by Alfred Schutz as well as Gurwitsch: the first stresses the metaphor of the projector, which sheds a strongly channeled light into the surrounding darkness, the second uses the concepts of *focus* and of thematic field.[8] In addition, Gurwitsch provides us with a remarkable reconstruction of Husserl's analysis of attention and of the light paradigm and makes it more concrete :

> The usual comparison of attention with an illuminating spotlight is endorsed by Husserl. "That which, in a specific sense, is paid attention to is found in the more or less bright beam, but it can also retreat into the half shadow and into full darkness . . . . This change in illumination does not change what appears as to its own sense-content, but brightness and darkness do modify its manner of appearance..." (SPP, p. 180).

But beyond this double delimitation of perceptual consciousness shared by these three authors, the structure of perception appears as a gradual continuum of light and shadow, endowed with various half-shades, which are so many more or less distant adumbrations and horizons. It follows that visibility is never complete: it is always mixed with obscure dimensions. In that respect, our perceptual activity is finite and limited.

Most of the time we act while being pushed and pulled by projects and goals: we do not wait for a complete reflection to have taken place in ourselves concerning the different risks of the action we are planning before undertaking it.

---

[7] Edmund Husserl, *Cartesianishe Meditationes*, 1950, *Hua I.*

[8] Alfred Schutz, *Der sinnhafte Aufbau der sozialen Welt*, Vienna, Springer, 1932 ; Aron Gurwitsch, *The Field of Consciousness.*

Thus, we gradually develop various know-hows, which adapt more to our reflection than they often anticipate it in addition to making it more mature, more concrete, and more embodied. By definition, our knowledge is never complete and our action builds upon knowledge in a partly blind way: it thus finds its clarification in the very development of the given action.

Heidegger emphasizes the finitude that mainly characterizes our perceptual activity in *Sein und Zeit.*[9] This is its limited elucidation through theory: the unfolding of our action is linked to an always open process of becoming visible, which means that it is never fully complete. If it was, we would be identified with an omnipresent and omniscient observer: such is the radical incompleteness of the dynamics of our interpretative activity. Explicitation corresponds to this indefinite process through which the various aspects of an identical phenomenon are brought to light in different ways. Following Heidegger's analysis, but with a more pragmatic stress, Hubert Dreyfus writes of the "*transparency*" of our know-hows,"[10] i.e., their familiar functional operativity, which is precisely deprived of any reflective clarification and could be paradoxically called their habitual opacity.

B) The absolute invisibility of our sensory experience. Such an immanent transparency makes our acting constitutively invisible to ourselves while it is being performed. The abyssmal depth of our sensibility consequently allows our body to escape us. Such an invisibility to ourselves has been described in different ways by phenomenologists:

1) Maurice Merleau-Ponty stressed the dimension of invisibility and the primacy of touch, which corresponds to the depth of our visibility and underlines the mobility of our sensory and emotional movements;[11]

2) Jacques Derrida questions the optocentrism of the philosophical tradition and follows Merleau-Ponty in advancing touch to the first place, i.e., what he calls the haptism of the bodies, which is nothing else than a blind inter-tactile incorporation;[12] and, finally,

3) Michel Henry thematizes the radical immanence of our self-affection, this pathetic invisibility of our flesh where it has the result that it is impossible to

---

[9] Martin Heidegger, *Sein und Zeit,* §§14-18.

[10] See Hubert L. Dreyfus, *Being-in-the-World : A Commentary on Heidegger's* Being and Time, MIT Press, Cambridge, 1991 ; *What Computer's Still can't do,* MIT Press, Cambridge, 1999. See also the Special Issue of *Phenomenology and the Cognitive Sciences*, Volume 1, No. 4, 2002 dedicated to "Hubert Dreyfus and Problem of Representation" and especially Dreyfus's and Clark's articles, respectively "Intelligence Without Representation—Merleau-Ponty's Critique of Mental Representation" and "Skills, Spills, and the Nature of Mindful Action."

[11] Maurice Merleau-Ponty, *Le visible et l'invisible* (Paris: Gallimard, 1966) and Susan Cataldi, *The Depth of Emotion (*Albany *SUNY* Press, Forthcoming).

[12] Jacques Derrida, *Le toucher* (Paris: Galilée, 2000).

reach the other in his or her flesh. Thus one makes love "with eyes shut," in the framework of an eroticism of opacity and blindness which has a tendency to self-eroticism:

> Such is in effect the consequence of the duality of instinctual movements each following its own trajectory, each ending with its own trajectory, which remains in itself, despite its intensity, leaving the other's pleasure in an inaccessible place. To grasp the phenomena in the strictness of their immanence, it would not be suitable to speak of eroticism but of autoeroticism.

Or, again, "desire fails to reach the other's pleasure where it reaches itself."[13]

C) Gurwitsch's "being together": the affective closure of fusion vs. *the ways of openness in partnership and membership.* It seems that immersion into immanent invisibility and into the tactual depth of perception leads to an affective radicalization that gives primacy to immediacy and does not allow a truly intersubjective context to arise. Because the affective and tactual depth of perception is an inner deepening, it also gives way to a closed-up perceptual activity of the subject. By contrast, the chiaroscuro of perception draws on the openness of our activity towards the world and paves the way for a possible "contextualized," situated and embodied intersubjectivity that follows from the very Gurwitschean notion of context as it is socially applied to community and society in the problematic of "consociates."

Gurwitsch does not emphasize his notion of "context" in *Human Encounters in the Social World*,[14] but it clearly underlies his analysis of "being together" as it is described in the third part of this book. Drawing on Scheler's account of the "milieu" and putting aside the argument of inference by analogy that he refers to Erich Becher, Gurwitsch emphasizes three different main modes of being together: 1) Partnership, 2) Membership, and 3) Fusion. Whereas the last mentioned involves the close-up experience of the other, partnership and membership are social relationships characterized by the affective mutual sharing of a common situation. Instead of "context," which is used much later, Gurwitsch prefers to use the term "situation"; the prevalence of a unitary feeling guides the related persons and produces a "fusional" being together, whereas partnership and membership are founded on a transitory relationship, which will never lead to a felt union or "oneness."

---

[13] Michel Henry, *L'essence de la manifestation*, Paris, P.U.F., 1963, §56; *Incarnation. Une philosophie de la chair*, p. 302 and p. 303.

[14] Aron Gurwitsch, *Human Encounters in the Social World* (1929-1930), Pittsburgh, Duquesne University Press, 1979. See also Shaun Gallagher's essay in the present volume.

The main difference between partnership and membership consists of the temporality of each relationship: while partnership is a punctual and local mutual sharing of an activity (Gurwitsch's example is of two men working together at cobbling a street, the first one laying the stones, the other one knocking them into place), membership corresponds to a historically sedimented time and involves not only two people but a whole group. Nonetheless, contextuality remains central in these two modes of intersubjective being together, being *hic et nunc* or sedimented in time and space: such modes of consociality leave open the possibility of distances (shadows) in relationships as much as genuine proximities (illuminations), whereas the third and fusional mode remains situated in the invisibility of a unifying affectivity.

With respect now to such an alternative between perceptual half-lighted openness and affectively invisible and fusional closure, some other phenomenologists ultimately insist on the experience of the superabundance of light, which is familiar in how the brightness of luminosity becomes denser, stronger, and more radiant, and may cause blindness. To what extent is such a pure luminosity able to open the way for a truly "other" intersubjective experience?

III. Vertical Context: The Radical Embodiment of a Relational Ethics

A) The superabundance of light: optical dazzling between blindness and wonder. Of course the radical invisibility of our sensuous movements is in sharp contrast with the experience of a flashing that corresponds to a pure luminosity where all unexplored sides are set off in shadow.

Following the model of an adequate perception, Husserl very early explicates the ideal possibility of complete visibility, which corresponds to an intuitive paradigm of truth. *Intueri* indeed means "to see," and the extension of organic seeing to the vision of *eidē* (*Wesenschau*) gives way to the possibility of an absolute seeing, compatible with the possible vision of essences and ideas.

More recently, Jean-Luc Marion has brought to light the descriptive possibility of a "saturated phenomenon,"[15] one which offers all the characters of an unseen phenomenon, insofar as, on the one hand, it goes beyond the limits of the givenness of the object in the very framework of its horizon and, on the other hand, it does not correspond to the very subjective mode of givenness of the ego. We have therefore to do with a phenomenon that the author calls a "revelation," which leaves absolutely nothing in the shadow, be they sides of the object or features of the subject. As it is light saturated, such a phenomenon is easily daz-

[15] Jean-Luc Marion, *Etant donné*, Paris, P.U.F., 1997.

zling and even blinding, which sometimes may lead one to identify it with the Transfiguration of Jesus Christ.

Furthermore and in contrast with Henry's invisible and solipsist autoeroticism, Marion attempts a different description of erotic intersubjectivity in his latest book, *Le phénomène érotique.*[16] It is generally founded on the experience of a full and absolute giving of oneself to the other and endeavors to break through the long-standing opposition between *eros* and *agapè*. Though it tries to go beyond the distinction between loving and loved, it preserves the distinction between two subjects, the one described as the loving one and the other described as the loved one: it is not the experience of a mutual erotic relationship but the experience of the look of a loving subject at a loved subject. In that respect the description of erotic intersubjectivity remains here wholly one-sided and insufficiently contextual.

B) The inner luminosity at the crossroads of phenomenology and mystical experience: the smiling face. But all these approaches to the luminosity of the lived body however illuminates it from an external source, be it in an excessive way, as a lack, or in its half-appearances: in each case it keeps an unsatisfactory distinction between the illuminated object and the illuminating subject, whereas the very notion of context in Gurwitsch paves the way for a holistic transcending of such a duality by advancing its very specific and open mutually intersubjective mediality. How is it possible to understand that a lived body may non-dualistically receive its light from itself, i.e., that it may produce it from itself?

At the very moment when we endeavor to keep up the latter with the scope of the phenomena, which is obvious when habitually observing them, it is necessary to carry the phenomenological analysis further and have it reach the threshold of a mystical description.

Beyond the static Husserlian approach proper to visual perception, which leaves the dual distinction between subject and object open, genetic phenomenology can combine with the Gurwitschean holistic contextual move in order to provide us with a dynamic scheme. In that respect, both account convergently for the luminosity that our lived body bears and which irradiates from it: 1) because of the inner strength of its sensuous movements (kinetic and proprioceptive), it is endowed with its own energy, which furnishes it with an unequaled creative power; 2) such a productive strength makes of the lived body the primordial earth of vision and visibility, the burning source from which come and proceed the beams of its luminosity. In that respect, the beam from the eye—the look—is a condition of vision without ever being able to see itself: the incandescent self-igniting capacity of the body is thus added to its power; and 3), as an originary and archaic locus of luminous strength, the lived body is consequently the first point of *orientation* for the other spatial bodies: it serves as a guide, it

[16] Jean-Luc Marion, *Le phénomène érotique,* Paris, Fayard, 2003.

structures the latter, ties together the different beams which radiate from the other bodies.

Such a description of the immanent procession of light from the inner source of the body, from the genesis of the bodily radiating, is fully convergent with the mystical experience of illumination, which in turn coincides structurally with the non-dualistic open notion of mutual context in Gurwitsch. As the lived body (*Leib*) is the dynamic pole, objectively unspatializable, of sensory moves which are so many luminous beams each of them being projected into space, in the same manner God is compared to the sun as an essence of energies, which are so many tangible participations in God without ever being equal to God, according to Palamas in his doctrine of divine energies during the XVth century.[17]

The phenomenal experience, which accounts both as exemplary and as everyday for such an inner source of luminosity, is the expression of the features of the face and, more particularly, of its smiling expression. As a phenomenologist, Emmanuel Levinas interestingly stresses the exposure of the face of the other to my look, which makes me vulnerable as I never was before, takes away every possibility of objectivation, and takes me back to my complete nudity. In this respect, the Levinasian face is, in his words, a pure "expression without context," which means that it is entirely contextual without any object, a sheer holistic form, and truly intersubjective: it is completely a givenness of love, an absolute self-giving, and it cannot be reduced to the sum of the features of a physiognomy, which can always be totalized and objectified.[18]

Such an expression of the absolute at the heart of the very finitude of a given face is nicely met in Olivier Clément's expression of an "internal visage." This amounts to saying that the face is the source of a beam that does not proceed at all from the other, but from the interiority of the spirit, i.e., from the goodness of the soul related to the courage of the heart.[19] What also accounts for the smile of praying people, e.g., the Holy Seraphim's from Sarov or Holy Silouane's, which is also called the "smile of the eyes."[20] Such a smile proceeds directly from the looking eyes and is not in the lips. It enlightens every feature so as to transform the being itself.

But the Levinasian face appears to be overexposed to the superabundant light of the Other (God), so that it does not seem that I am able myself to produce such an illumination. The inner luminosity we are in quest of is not relevant in an ethics founded on such an asymmetry between the Overluminate

---

[17] Grégoire Palamas, *Triades*, dans *La prière du cœu*, Bellefontaine, 1994, p. 21-22.
[18] E. Levinas, *Ethique et infini*, Paris, Librairie Arthème Fayard, 1982.
[19] Olivier Clément, *Le visage intérieur*, Paris, Stock, 1978, namely: "Le visage et l'icône," pp. 10-67.
[20] Holy Seraphim from Sarov; Holy Silouane; A. and R. Goettmann, *Prière de Jésus, Prière du cœur,* Paris, Dervy-Livres, 1982.

Other and the Enlightened I. In short, we have to do with an ethics of the Unmeasured, which means Absoluteness, and not with an ethics founded on relational mutuality. Such an absoluteness of the Other is also what leads to its tendential abstractness, because it finally results in something separated from me as a finite lived body. In that respect it is difficult to see how such an Other can actually be radically embodied.

In contrast with the Levinasian face, we would like to urge the smiling face (Levinas never says that the face is smiling), which is a concrete and embodied expression of its immediate relational nature. Insofar as my smiling expression cannot but be related to the other smiling at me, the smiling experience is an originarily relational experience. Even the experience of "inner smile" already discussed refers to the sedimented experience of a previous intense relationship we are recalling while smiling.

In accord with Gurwitsch's original experience of contextuality, what is today called "contextual therapy" (inspired by the situated and embodied Second Cybernetics)[21] is precisely founded on a genuine relational ethics where the other is not exposed to my violence, as it tends to be with Levinas, but is truly free. While "re-esteeming" the other by holding back, I give him to himself the possibility of existing freely.[22]

*

* *

To conclude, the possibility of a "vertical context" both emerging from its horizontality and its very depth and exceeding it from within in the unique experience of the overluminous smiling face brings about a renewed concept of intersubjectivity defined by its ethical dynamics of self-transcendence. It helps shed a renewed light on the Gurwitschean notion of an organic non-dual intrinsically contextual experience.

---

[21] Ivan Boszormenyi-Nagy, *Fundation of Contextual Therapy,* Brunner Mazel, 1987; P. Michard et G. Shams Ajili, L'approche contextuelle, Paris, Ed. Morisset, 1996; Mony Elkaïm (éd.), *Panorama des thérapies familiales*, Paris, Le Seuil, 1995; P. Michard, " De l'éthique intime," *Groupe familial*, n°133, 1991.

[22] Cristos Yannaras, *La liberté de la morale*, Genève, Labor et Fides, 1987.

Chapter 3

Schizophrenia: A Disturbance of the Thematic Field

Louis A. Sass
Rutgers University

Abstract

This chapter explores the relevancy of Aron Gurwitsch's phenomenological analysis of the field of consciousness—in particular his distinction between theme, thematic field, and margin of awareness—for understanding certain abnormalities of consciousness and cognition in schizophrenia. Alterations in the organization of the field of consciousness have long been recognized as a key feature of the schizophrenic condition. It has been difficult, however, to specify these aberrations very precisely, or to capture the nature of schizophrenic abnormalities in contrast with vaguely analogous abnormalities that are found in mania, dementia, and various neurological conditions. Traditional theories of attentional dysfunction have been found to be inadequate. I argue that (A) the distinctively schizophrenic alterations of perception, thought, and language demonstrated in recent research pertain especially to what Gurwitsch calls the "thematic field" and (B) disturbances involving the margin or the theme may best be viewed as secondary consequences of this disturbance of the thematic field. I discuss a number of abnormalities of schizophrenic cognition, including perplexity, dissemination (proliferation of meanings), fragmentation, perspectival slippage, and certain quasi-surrealistic ways of experiencing what is familiar or strange.

Introduction[1]

Alterations in the organization of the field of consciousness have long been recognized as a key feature of the schizophrenic condition. It has been difficult, however, to specify these alterations very precisely, or to capture the nature of schizophrenic abnormalities in contrast with vaguely analogous abnormalities

---

[1] For helpful comments on this paper, I thank Jeff Yoshimi, Jeffery Geller, John Barresi, Steven Silverstein, and Lester Embree.

L. Embree (ed.), Gurwitsch's Relevancy for Cognitive Science, 59-78.

found in mania, dementia, and various neurological conditions (Hamilton 1976, p. 162). In this essay I will explore the relevancy of Aron Gurwitsch's phenomenological analysis of the field of consciousness for understanding this aspect of schizophrenia. I argue that the distinctively schizophrenic disruptions of the organization of perception, language, and thought (the "disorganization" syndrome; Liddle, Carpenter, Crow 1994) pertain especially to what he calls the "thematic field," and that disturbances involving what he calls the margin or the theme may, in many cases, best be viewed as secondary consequences of this disturbance of the thematic "field."

It would be too much to argue that Gurwitsch's phenomenological account offers a dramatic conceptual breakthrough for schizophrenia research—as we shall see, many of his key points have been developed independently by psychologists or psychiatrists working in other traditions. His phenomenology does, however, offer a useful conceptual analysis and vocabulary for specifying schizophrenic abnormalities of consciousness and cognition. It should help to organize and orient our understanding of this enigmatic disturbance of context.

I. Gurwitsch's *The Field of Consciousness*

The central purpose of Gurwitsch's *The Field of Consciousness* (FC) is to explore the kinds of relationships that link together the various elements and objects of human acts of awareness. He describes his book as "an analysis of context in general [with] eventual disclosure of different types of context" (FC, 2). His phenomenological investigation reveals a crucial heterogeneity in the nature of these relationships—a discovery that is at odds with influential psychological theorists who, for all their differences on other issues, nevertheless share a homogenizing tendency to assimilate all mental relationships to a single type. For Hume, British empiricism, and the associationistic tradition in general, the field of awareness is an aggregate of intrinsically unrelated elements bound together by "external" factors. By contrast, for Henri Bergson and William James, all mental states composing consciousness reveal a "thoroughgoing interpenetration and intrinsic interconnection" (FC, 3). In Gurwitsch's view, both these classic positions are oversimplified. He argues that there are, in fact, three domains within the field of consciousness, each with its own, distinct principle of organization.

Gurwitsch's first domain is that of what he calls the "theme." The theme is that which engrosses the subject, stands in the central focus of his attention, and "upon which his mental activity concentrates" (FC, 4, 55). The second domain, the "margin," is the realm of objects of awareness that are simultaneously present to consciousness but that are not experienced as having any relevancy to the focal theme. The margin includes elements or aspects of three orders of existence: one's ongoing bodily existence, the perceptual surround, and "a segment of the stream of conscious life." The third domain—a kind of middle realm—is that of the thematic field, defined by him as "the totality of facts, co-present with the theme, which are experienced as having material relevancy or pertinence to the theme" (FC, 55). He describes the theme—thematic-field—margin structure as "a formal invariant of organization having universal significance" (FC, 379); as such, it should apply equally to fields of consciousness involving perception, thinking, language, and memory.

In Gurwitsch's view, a certain type of relationship is characteristic of each of these three realms. The *theme* is characterized by the kinds of relationships that the Gestalt psychologists found in basic perceptual figures. Here each constituent "exists in the very qualifications by which it is defined and made to be that which it is in a given case, only in conjunction with, and determined by, co-constituents" (FC, 139, 358). The "parts" of the theme are *not* independent and self-contained. They manifest the essence of Gestalt-conjunctions or "Gestalt-contextures" (FC, 358), which is "the equilibrated co-existence of mutually dependent constituents" (FC, 139); this gives them the kind of coherence that allows them to stand out from the wider background as a discrete and coherent whole.

Marginal relationships (i.e., relationships between what exists in the margin and in either the theme or the thematic field) combine temporal co-presence with irrelevancy. Gurwitsch explains: "[C]onjunctions between both the theme and the thematic field, on the one hand, and on the other, the margin, are of mere temporal nature. They consist in either the simultaneous or immediately successive occurrence of the respective acts" (FC, 353).[2]

Relationships among aspects or elements of the *thematic field*, or between the thematic field and the theme, contrast with the Gestalt-type connections that organize components *within* the thematic focus. Elements of the thematic field are *not* "whole-parts" united by "Gestalt-coherence": here each constituent does have some kind of core identity that is independent of the other constituents. But in contrast with the elements of *marginal* awareness, the constituents of the thematic field hang together in a mutually implicative way, and they have relevancy for the focal theme. Thus the *way* in which each of these constituents is experienced is a function of some larger unifying significance, a significance that also determines the *aspect* of the theme that will emerge as significant. Gurwitsch speaks here of "unity by relevancy," which is bound up (as both source and product) with the unity of *context* (FC, 10, 341). The subjective correlate of the thematic field is the "attitude" of the subject or perceiver; "attitude"

---

[2] Gurwitsch does not deny that relationships with*in* the margin may be structured (see Gurwitsch 1985). A combination of mutual irrelevancy and mere co-presence is the essential feature of the relationship *between* margin and theme or *between* margin and thematic field; presumably, it may or may not apply with*in* the margin.

is bound up with the "perspective" within which something is seen or known (FC, 322, 380; Gurwitsch 1966, pp. 201, 203, 224). Gurwitsch states that this middle realm—"thematic field"—is profoundly important in determining the "tinge" and significance of the theme. It determines the "*perspective* under which, the *light and orientation* in which, the *point of view* from which" the theme appears to consciousness and, with this, the theme's "positional index," which is the theme's position within the thematic field (FC, 359, 361).

A further aspect of the thematic field that would seem to be implicit in Gurwitsch's analysis is its role in determining the margin. If the margin includes whatever appears yet is nevertheless irrelevant, then *its* contents are also delineated—albeit in a negative sense—by the nature of the thematic focus and field. This would seem to imply that, when the thematic field has a looser, more diffuse, or more inconsistent quality, the very distinction between thematic field and margin might be affected. This could have several possible consequences. One possibility is that elements that would normally remain in the margin would be more likely to intrude into the focus, where they would appear as alien elements. (Gurwitsch speaks of the "incipient encroachment of an alien system upon that in which [one is] living at present"; FC, 384.) A second possibility is that the very definition or significance of the thematic field would be affected, with marginal intrusions having the kaleidoscopic effect of *shifting* the thematic field itself toward some entirely different definition or perspective.

There is at least one additional potential consequence of a weakened or fluid thematic field—not considered in Gurwitsch's *The Field of Consciousness*—that is relevant for present purposes. This concerns the impact that a loss of overall perspective, point of view, or orientation might have on the Gestalt stability and wholeness of the *theme*.

Just now I have described how the absence or weakening of an organizing and orienting perspective (i.e., a stable thematic field) might lead to a permeability or slippage of this orienting field. In this sense it would be conducive to or otherwise associated with a mode of attention that is easily intruded upon, or with an attitude that drifts or meanders in aimless fashion. This weakening of orienting perspective might, however, also be associated with attitudes or forms of attention that are, in a sense, rigid or hyper-focused—where the normal, practically-oriented movement from aspect to aspect of a thing, usually guided by an orienting perspective of "care" or "concern" (see Heidegger 1962, re the role of *Sorge* or *Besorgen*) is replaced by a fixated stare that can, in a sense, bore through, fragment, and de-realize its focal or thematic object. As a result, the Gestalt unity of the thematic focus would be broken up into parts. Under these conditions, the larger unifying patterns—especially those that require an organizing sense of "concern"—would tend to disappear. The field of consciousness would not dissolve into a complete chaos or a blooming, buzzing confusion but, instead, would resolve into a loosely united field of smaller but more intrinsically salient Gestalts that seem to impose themselves on the subject's awareness.

It should not, then, be surprising to discover some degree either of marginal intrusion or of Gestalt breakdown even in cases where the core problem concerns disturbance of the thematic field. It should, however, be possible to show that the intrusions and breakdown are less prominent than contextual disturbance, and that they have a secondary status in relation to the latter.

Summary: Let me sum up by stating what predictions one might make, based on the above analysis, about the likely consequences for overall cognitive functioning of a disturbance of the thematic field. First, of course, one would expect that the most prominent alteration of consciousness would be a diminishment of the sense of organizing context, orientation, or perspective. Patients with this disturbance should have more trouble grasping the overall context or meaning of situations or stimulus fields than with the accurate perception of individual details. Also, although one should expect the occurrence of intrusions of irrelevant stimuli or thoughts from the marginal domain, this is unlikely to constitute the primary or dominant disruption of awareness and orientation. And when such intrusions *do* occur, they are unlikely to invade an otherwise robust and focused theme/thematic field but, instead, will be preceded by a weakened or unstable ongoing orientation and sense of context.

Finally, one would expect a preservation of the fundamental capacity for Gestalt perception, particularly when this concerns perception of basic, geometrically defined visual shapes and the like in other sensory domains. The perception of less obvious *Gestalten* or larger wholes that are dependent on an ongoing sense of context and relevancy, would be more likely to be compromised. But since this former tendency would be a *consequence* of the loss of contextual relevancy, it should vary in accordance with the latter. More blatant disruptions of normally strong *Gestalten* may sometimes occur. But if so, this is likely to happen not as a passive occurrence but in the presence of an abnormally fixed and narrowly focused form of attention (a kind of "stare") that seems to fragment the larger whole precisely by picking out and framing distinct components within it.[3]

---

[3] It can be difficult (perhaps more difficult than Gurwitsch implies) to draw a line between margin and thematic field. Consider one of Gurwitsch's examples of marginal awareness: the theoretical scientist's awareness of his own bodily posture while thinking about some scientific theme. Body-posture and scientific problem may at first seem mutually irrelevant, merely co-present. Closer analysis may well suggest, however, that background awareness of postural and kinesthetic dimensions of the lived-body often provides a kind of isomorphic medium that supports or symbolizes the problem domain (see Merleau-Ponty 1962). This suggests that there can be a kind of bleeding together of margin and thematic field.

II. Earlier Research on Attention and Perception in Schizophrenia

In the 1960s and 1970s, the dominant theories of schizophrenic cognition hypothesized a defect or deficiency of "selective attention," that is, of the ability to exclude irrelevant sensory input from conscious awareness during the act of perception. The forms of experience posited by this hypothesis of a defective attentional "filter" are, essentially, states of distraction, confusion, and sensory overload. McGhie and Chapman (1961) speak of "an undifferentiated mass of incoming sensory data" or "involuntary tide of impressions" that sweep over "an undifferentiated protoplasmic consciousness" (pp. 110, 112). This classic account emphasizes intrusion into consciousness of ("marginal") material that, in the normal mind, would be classified as irrelevant and excluded from focal or thematic awareness. In contrast with Gurwitsch's tripartite model, McGhie and Chapman assume a dualistic framework, with potential contents of consciousness divided into the categories of the relevant and the irrelevant. The hypothesis of a defective attentional filter has a rather mechanistic cast: The early perceptual anomalies are interpreted as rather straightforward results of *failures* or *defects* of cognitive functioning, with the patient playing an almost entirely passive role.

The defective-attentional-filter account, once the dominant theory of schizophrenic cognition, has been largely rejected in the last two decades or so (Cutting 1985, pp. 208, 212f, 216; Heinrichs 2001, p. 36). Many researchers now recognize that the disturbances found in schizophrenia patients are, in fact, less widespread, persistent, and debilitating than this theory would predict. Research demonstrates, e.g., that schizophrenic "distractibility" itself is, in some respects, far more selective than this rather mechanistic theory would imply: Compared to normal individuals, schizophrenia patients were found to be more easily distracted by tape-recorded voices *only* when these voices were speaking of topics related to the patients's delusions (Cutting 1985, p. 213). Also, schizophrenics seem *less* likely to be distracted by outer than by inner stimuli—that is, by the normally unnoticed workings of their own minds (Frith 1979). In his survey of the psychological research, Cutting (1985, pp. 208, 212f, 216) concluded that distractibility or disturbance of concentration is at least as common in depression and in psychotic disorders in general as in schizophrenia (see also Hemsley 1976; and references in Frith & Allen 1988, p. 176). Cutting (1985, p. 214) does however note that schizophrenia patients often have trouble *maintaining* a focus of attention.

The defective-attentional-filter hypothesis would seem to predict only deficits of cognitive and perceptual performance. However, on tasks where grouping or configuring of stimuli would interfere with performance, patents with schizophrenia often perform *better* than normals do (Silverstein et al 2000; Cutting 1985, chap. 14). This pattern of deficits and strengths conflicts with the defective filtering model (and also with models hypothesizing slowed processing and reduced capacity for informational input), but is consistent with hypotheses concerning abnormal "cognitive" or "contextual coordination" or "perceptual organization" (Knight 1992).

I would add that the attentional-filter hypothesis does not seem to capture certain key features of subjective experience in schizophrenia. It would seem to predict a real chaos—something like the "bloomin,' buzzin' confusion" of which William James famously spoke, but this is not what schizophrenia patients actually report. The defective attentional filter account is certainly not consistent with the uncanny, alienated, often static and surreal qualities of the world of the schizophrenic prodromal phase. Nor does it capture certain associating feelings of hyper-intentionality, of a kind of anxious *directedness* of attention that will be discussed more below.

An alternative account of the abnormalities of schizophrenic awareness—less well-known in Anglophone psychiatry—has been inspired by the theoretical approach of Gestalt psychology. According to the German psychiatrists Klaus Conrad (1958) and Paul Matussek (1987; Hamilton 1976, pp. 157-162), writing in the 1950s, the fragmented or disorganized quality of the schizophrenic's perceptual world is a direct manifestation of a breakdown or "change in the Gestalt structure of the perceptual world" (Matussek 1987, p. 100). Emphasis on Gestalt breakdown as a key factor in schizophrenia has been criticized on the grounds that, on various standard tests of perceptual ability, schizophrenia patients do not, in fact, typically manifest the disturbances of basic Gestalt that are found in many patients with organic brain damage (Heinrichs 2001, pp. 30ff; Silverstein et al 1998).[4]

III. Recent Research on Attention and Perception in Schizophrenia

There is no doubt that attentional processes are often abnormal in schizophrenia, and that distractibility does occur. It is also true that schizophrenic individuals sometimes experience a fragmentation of larger organizational qualities of experience. The strongest recent research suggests, however, that these

[4] Actually, Conrad's and Matussek's accounts are less purely Gestaltist—and more congruent with Gurwitsch—than they initially appear. Gurwitsch (1964) sometimes refers to "two types of wholes of Gestaltist nature," the one involving unity by Gestalt-coherence, the other unity by relevancy (p. 365), the first pertaining to the theme, the second to the thematic field. Although Conrad and Matussek do not mention this distinction (to my knowledge, at least), their examples of what they call "change in the Gestalt structure" nearly always emphasize unity by relevancy.

abnormalities are best understood not as manifestations of deficient attentional filtering or a general broadening of attention (of a generalized inability to focus and exclude) nor as a loss of basic Gestalt perception, but, rather, as consequences of some kind of disturbance in the ongoing sense of context.

Schizophrenia does not involve the total breakdown of significant form, nor the blooming buzzing confusion, that would be predicted by a breakdown of the basic capacity for Gestalt-perception or by an incapacity to exclude the irrelevant. The central factor seems, rather, to be a difficulty in determining, at any given moment and in a consistent way over time, what should *count* as relevant.

One recent theory of schizophrenic cognition, e.g., postulates "a reduced influence of previously acquired information of a kind that would normally direct selective attention" (Jones, Helmsley, Gray 1991, p. 415). Others speak of the role of abnormalities of "contextual coordination" or "perceptual organization" in schizophrenia (Phillips & Silverstein in press; Knight 1984). This disturbance of context-processing seems to be bound up with an unusual perceptual strategy or manner of *allocating* attention (Cutting 1985), which, in turn, is connected with the schizophrenia patient's unusual, often unstable set of interests and an eccentric, often impractical orientation to the world (Zubin 1975). The earlier, selective-attention hypothesis emphasized the role of marginal intrusions. By contrast, the more recent theories emphasize disturbances in what Gurwitsch would call the thematic field—with the marginal intrusions (disturbances of attentional filtering or selective attention) and instances of fragmented perception (Gestalt breakdown) that *do* occur being considered secondary to a more primary disturbance of the ongoing and organizing sense of contextual significance or meaning.

An exciting feature of this recent work on context-disturbance is that a single abnormality seems able to account for a wide variety of cognitive abnormalities found in schizophrenia. What has been called "contextual coordination," "contextual processing," or the "coordination of contextually related stimuli" operates across domains in implementing such processes as selective attention, perceptual grouping in vision (perceptual organization), and lexical disambiguation in language production and comprehension (Silverstein, Kovacs, Cory, Valone 2000, p. 12). It suggests "commonalities between the way that visual context structures object representations in vision, and the way that meaning serves as a context to structure linguistic and conceptual representations" (p. 18). A disturbance of context processing could therefore provide a unifying explanation of the disturbances shown by schizophrenia patients across a variety of domains including attention, perception, thought, memory, and language (Silverstein & Shekel 1997; Servan-Schreiber, Cohen, Sternward 1996, p. 1105). As we shall see, it may also be relevant for understanding disturbances that concern the coordination of both simultaneous and successive forms of information or sensory input.

In this section, I summarize the main findings of recent laboratory studies, and also offer a few relevant observations along with illustrative quotations from patients.

A) Context processing: On various tasks, the performance of schizophrenia patients can be accounted for by a reduced ability to utilize contextual information (Cohen et al, 1992; Shakow 1962). In a study by Cohen, Brach, Carter, and Servan-Schreiber (1999), for example, subjects observing a visual display of a sequence of letters were instructed to press a button only when they saw the letter X, and only when X was preceded by an A. Subjects with schizophrenia or schizoaffective illness did especially poorly on this task—leading the authors to conclude that these subjects demonstrated "a degradation in the processing of context needed to select task-appropriate action" (p. 132). (A number of experimental manipulations and controls were used to assure that the findings were not secondary to a generalized cognitive deficit.) These authors use "context-information" to refer to "information that must be actively held in mind in such a form that it can be used to mediate task appropriate behavior" (p. 120).

The inability to maintain an organizing context is perhaps the main source of the characteristic schizophrenic *perplexity* that has long been identified (Störring 1987). Describing the above-mentioned experiment in Gurwitschean terms, one would say that at the moment an X is perceived and responded to, it constitutes the theme or thematic focus, while the A—together with the background conditions of the game (e.g., pushing the button)—constitutes the thematic field. The relationship between A and X in this experiment is obviously not a relationship of Gestalt-coherence: notice that the A and X are not constituents of an indivisible whole; each has its independent identity. Nor is it a relationship of mere copresence: although A's significance for interpreting or classifying X is arbitrarily imposed by the rules of the game, this does not make it any the less a matter of a *relevancy* relationship.

In another study, Silverstein, Matteson, and Knight (1996) showed that whereas selective attention processes based on purely physical characteristics of auditory stimuli (numbers and animal or animal-like sounds) were intact in schizophrenic subjects, these same subjects (unlike normal controls) did not use higher level, contextual information to reduce the interfering impact of a stimulus that, according to this information, should have been classified as irrelevant. Apparently, such a failure is not restricted to the auditory realm. In a study using visual stimuli, Carr, Dewis & Lewin (1998) found no evidence that schizophrenia is associated with a "preattentive abnormality in visual search using stimuli that differ on the basis of physical characteristics" (p. 69). Their findings were more consistent with the possibility that "preattentive processing abnormalities in schizophrenia may occur only under conditions involving selected 'top-down' factors such as context and meaning" (p. 151).

Disturbed contextual processing has also been demonstrated in the linguistic performance of schizophrenia patients. Kuperberg, McGuire, and David (1998) asked subjects to detect a target word, such as "car"; unlike normal persons, schizophrenia patients were not distracted by linguistically abnormal contexts (e.g., "He ate the car quickly"), which suggests that they were not sensitive to context.

Consider also clanging responses—responses where a person jumps from word to word based purely on *sound* qualities of the signifier or sound vehicle. Given the arbitrary nature of the relationship between signifier and signified, the sound-vehicle would normally have the qualities of Gurwitsch's "margin": co-presence along with irrelevancy. (I am speaking here of normal, practical uses of language, not of poetry or song.) The train of associations in a clanging response could therefore be described as being driven by intrusions from what should have remained in the margin. Clanging sometimes does occur in schizophrenia (and also in mania; Solovay, Shenton & Holzman 1987, p. 20)—a finding that is consistent with my hypotheses stated above. It seems however, to be less common than in certain forms of aphasia (Peuser 1987, p. 243). More characteristic of schizophrenia in particular are experiences in which the patient becomes aware of a proliferation of meanings generated by a particular word, phrase, or object (Lecours & Vanier-Clement 1976; Peuser 1987, p. 243)—what Derrida (1981) would call *dissemination.* As one patient explained,

> My trouble is that I've got too many thoughts. You might think about something, let's say that ashtray and just think, oh! yes, that's for putting my cigarette in, but I would think of it and then I would think of a dozen different things connected with it at the same time.

Another said:

> It's not that I can't concentrate right, it's just that I can't concentrate on the major issues. I get fogged up with all the different bits and lose the important things in the picture. (McGhie & Chapman 1961, p. 108)

A visual equivalent to this proliferation or fluidity often occurs on the Rorschach inkblot test, when a part of the inkblot may seem to transform or rotate perspectives before the patient's eyes. One patient, e.g., could not decide whether he saw poodles or ladies; another described "two people … one minute this appears like their eyes and the next this appears like their entire body holding on" (Johnston & Holzman, pp. 93f). In the so-called "contamination" percept, pathognomonic of schizophrenia, two perspectives appear to be present simultaneously, as if overlaid in a double exposure. One patient saw the whole inkblot as both a butterfly and the world: hence "a butterfly holding the world together" (p. 99). These are instances of "fluid thinking," which research shows to be especially characteristic of schizophrenic abnormalities of cognition, as opposed to "irrelevant intrusions," which are more common in mania (Holzman, Shenton & Solovay 1986, p. 368).

I would interpret such instances of semiotic or perceptual slippage as consequences not of marginal intrusions, but of an absence of the disambiguating semantic fixation or "positional index" that is normally provided by the thematic field (FC, 361). The orienting perspective taken toward an object of thematic concern, which normally anchors the meaning of ambiguous words and things, seems to have come unmoored. Given the ubiquity of potential ambiguities in language and experience, this weakening of determining context is likely to have a widespread, unhinging effect on both language and thought (see Derrida 1981; Sass 1992, pp. 198-205).

Schizophrenic discourse and trains of thought often demonstrate primarily a meandering or tangential quality, rather than the true distractibility more common in mania.[5] The loosening of thematic field allows thought to *wander*; but this wandering seems to be generated, in some sense, from *within*, rather than in response to intruding mental contents or to stimuli beckoning from somewhere entirely outside the field of concern. Semantic ambiguities, necessary qualifications, and incidental but not irrelevant details open up alternative, tangential pathways; these tangents are followed until the original track is practically forgotten.

B) Fragmentation: We see, then, that recent empirical research on schizophrenia does not favor the attentional-filter-breakdown theory, at least as this has traditionally been formulated. Recent research also tends to be inconsistent with the hypothesis of a breakdown of basic Gestalt processing in schizophrenia. A study by Silverstein, Osborn, West, & Knight (1998; also Chey &

---

[5] I am suggesting that mania is more often characterized by margin-to-theme transitions and schizophrenia by rather uncontrolled, kaleidoscopic kinds of thematic-field-to-theme transitions that are superimposed on an unusual mutability of the thematic field. The difference I am hypothesizing is not absolute, by the way, only a difference of degree; presumably, schizoaffective thinking would be more equally characterized by *both* kinds of transition. (See Shenton, Solovay & Holzman 1987, for data consistent with this hypothesis.)

Yoshimi (this volume) describes the two, just-mentioned Gurwitschean types of "field transition." He goes on to characterize thematic-field-to-theme transitions as less "abrupt and "jarring' than are margin-to-theme transitions. Yoshimi is probably thinking of the kinds of thematic-field-to-theme transitions that occur in the context of a relatively stable thematic field. The lack of stability of thematic field in schizophrenia can make this type of transition very abrupt and jarring in such patients. (I am grateful to Jeff Yoshimi for helpful conversation on related issues in October 2002.)

Holzman 1997) suggests that, by contrast with certain neurological patients, neither acute nor chronic schizophrenia patients show any deficit in their ability to process stimuli with strong Gestalt features—i.e., highly configural stimuli with continuous or nearly continuous contours. It is only with more complex or ambiguous Gestalts, or with "weakly grouped stimuli," that schizophrenia patients show failures of perceptual organization (Knight & Silverstein 1998). Context processing (Gurwitsch's "thematic field") is crucial for perceiving complex or ambiguous forms. Rather than simply imposing themselves, such forms will only be perceived by a person who brings expectations or schemas to bear in his confrontation with the world; such expectations and schemas are weak or absent in a perceiver who fails to contextualize.[6]

Context and meaning tend to be holistic factors; their weakening or fragmentation in schizophrenia may therefore be related to the over-focused or analytic attentional style that has been found in schizophrenia patients as well as in many persons with schizoid and schizotypal traits (Goodarzi, Wykes, & Hemsley 2000). Some experimental studies have shown that, in schizophrenia, a larger, integrated Gestalt, such as a human face, may not be dominant over its component parts (Cutting 1985, p. 294; Frith et al., 1983). In his comprehensive review of psychological studies of schizophrenia, John Cutting describes the schizophrenia patient as "analytic and over-preoccupied with detail at the expense of gestalt or seeing order in things" (p. 305). He sees this as a "disorder of strategy rather than a failure of one aspect of perception."

A passage from *Autobiography of a Schizophrenic Girl* (Sechehaye 1970) gives a vivid sense of what this experience of fragmentation can be like: "I saw the individual features of [my therapist's] face, separated from each other: the teeth, then the nose, then the cheeks, then one eye and the other … [an experience that] prevented my recognizing her even though I knew who she was" (p. 37). Another patient's response to a human face seen on a Thematic Apperception Test card illustrates the distorting effect that this kind of over-focused attention—combined with indifference to common sense—can have. Most people simply take for granted that the expression in each of a person's two eyes is the same—that is, that the eyes, in this important sense, form a single unit that is expressive of a unified consciousness or perspective on the world. "Henry's" scrutinizing, however, led him to see each of a man's two eyes as having a different look: "It seems that the one [i.e., the eye] on the left is fatigued and the one on the right is worried and looks bitter" (Sass 1992, p. 459).

---

[6] Knight & Silverstein (1998) describe schizophrenia patients with poor premorbid social capacity as failing "to benefit from past experience and current context in detecting stimulus relations in complex environments and consequently in transforming sensory data into appropriate conceptual level representations and accurately assessing the behavioral significance of current stimuli" (p. 271).

Persons in the schizophrenia spectrum often seem to move in on the stimulus field in the sense of engaging in a kind of fixed, penetrating, over-focused stare that dissolves the more commonly recognized Gestalts in favor of their component parts (Goodarzi, Wykes, & Hemsley 2000). Thus liberated from the armature of familiar context, the parts may take on unexpected or unusual meanings. Staring does seem able to dissolve context (see FC, 291 re staring), but it seems likely as well that staring would be inspired or facilitated by a prior loss of context. Possibly the staring and the dissolution exist in some intimate kind of complementarity in which neither has overall priority over the other. There is, in any case, something paradoxical about the *experience* of this stare: although the staring often has a hyper-intentional, effortful, even aggressive quality, it may also feel, to the patient, like something that imposes itself on him and from which he cannot desist (see Sass 1992, pp. 68-74).

C) The Familiar and the Strange: So far our consideration of weakened context-processing in schizophrenia has emphasized concurrent or synchronic aspects, e.g., the role of simultaneous features of the surrounding spatial context. Context can, however, also be conceived in a diachronic (temporal, extended over time) way, that is, as involving the memory of preceding events and the expectations these typically generate (Phillips & Silverstein in press; FC, 97, 247, 280).

The diachronic aspect of the context disturbance reveals itself in responses to both the familiar and the strange. In 1991, Gray et al. described what they saw as a key disturbance of the "comparator" system in schizophrenia, namely, a failure to use previously acquired information to direct attention toward novel rather than familiar aspects of the world. Patients failed to inhibit orienting and startle responses toward stimuli that had been previously perceived, thus displaying a propensity to respond to the familiar as if it were novel. But along with this well-documented tendency, patients with schizophrenia also manifest just the opposite one. Research shows that schizophrenics are often *better* than normals at recognizing what should be an unexpected scene (e.g., a picture of a man holding a paint brush between his toes), as if they considered something bizarre to be quite unsurprising and probable (Polyakov 1969). Schizophrenics may fail to be surprised by strange or novel situations, but they can also be startled by what would seem to be perfectly normal;[7] the everyday may be experienced as shocking, but the unusual or the weird may be completely taken in stride (Sass 1992, p. 127).

There is something surrealistic about the subjective aspect of both these tendencies. If the first is illustrated by the paintings of Giorgio de Chirico—uncanny cityscapes imbued with an inexplicable strangeness—the second is

[7] Shakow (1963), quoted in Cutting (1985, p. 59): "A schizophrenic reacts to old situations as if new, and to new as if old."

illustrated by scenes from Luis Buñuel's surrealistic film "*Le Chien Andalou*"—e.g., a deadpan presentation of robed priests shown pulling a grand piano that bears the carcass of a donkey or a horse. Both tendencies seem to be captured by the concept of a disturbance of "probability prognosis" (Polyakov 1969). Normal persons approach the world with certain reasonable anticipations that direct their attention and awareness; their anticipations are not "empty expectancies, expectancies of anything whatsoever, but rather … of something which, however indefinite, fits into a certain pattern" (FC, 281). By contrast, persons with schizophrenia seem to enter each situation as if absolutely anything were not only possible, but perfectly plausible. This failure to impose expectations concerning the probable, or to ignore the implausible, is a main source of what has been called the "pathological freedom" (Sass 1992, p. 127) of schizophrenic thinking; it is a manifestation of the weakening of the thematic field.

IV. Perspectival Slippage

Instability or impoverishment of the thematic field is a source of the profound "perspectival slippage" that often characterizes schizophrenia. I use the latter term to refer not merely to a loose or shifting connection between particular mental contents (the essence of Bleuler's "loose associations" and of contemporary, neo-Bleulerian associationistic approaches, e.g., Spitzer 1997), but to a more global, kaleidoscopic shifting from one perspective or frame of reference to another. In *The Order of Things*, Michel Foucault (1994) draws a useful distinction between two types of incoherence that he refers to as "incongruity" and the "heteroclite" (p. xvii). Whereas the first is exemplified by the famous surrealist line—"beautiful as the chance encounter … of a sewing machine and an umbrella"—the second is exemplified by Jorge Luis Borges's imaginary encyclopedia in which animals are divided into a set of mutually incomprehensible categories or category systems ("tame, belonging to the emperor, that from a long way off look like flies," etc. [p. xv; see also Sass 1992, p. 120]).

Forty years ago the psychologist Andras Angyal (1964) described schizophrenic thinking as "not impaired so far as the apprehending of relationships is concerned." When the schizophrenia patient "fails in the solution of an intellectual task," writes Angyal, he "fails in the apprehending of system-connections"; he "drift[s] from one semantic stratum into another." Angyal defines a "relationship" as involving the pairing of two items; by contrast, a "system" involves an indefinite number of constituents arranged "according to some unitary plan with a 'dimensional medium,'" "context," "organized field," or "universe of discourse" (pp. 119-120). Gurwitsch uses the phrase "order of existence," and argues that this should be understood as the indefinite extension of a given thematic field or an indefinite continuation of context (FC, 381).

The schizophrenia patient prone to thought disorder often seems aware of a vast number of possibilities; but instead of focusing on one (such as color or shape on an object-sorting task), it is as if he takes "all the possibilities into simultaneous consideration," thus leading to a competition among inconsistent and incompatible *modes* of response. The psychologists Rapaport, Gill, and Schafer (1968, p. 475) speak of the schizophrenic's loss of "solidity and coherence." Because of the weakening or absence of an orienting thematic field, such a person's "relation to internal and external reality [is] altered from one response to the next [and] there is a new definition of the test situation and of his role in life with each response" (Schafer 1954, p. 107).

On the Rorschach test, e.g., the patient with schizophrenia may not stay within a single orientation or "set"—which is normally that of finding objects or people in the inkblots—but may, in heteroclite fashion, slip from describing such objects to discussing the look of the ink on the cardboard, his own style of responding, or even the nature of perception itself (Sass 1992, pp. 164-170), thus veering among what Gurwitsch (pp. 382 f) would call distinct "constitutive relevancy-principles." A young artist who had experienced a number of schizophrenic episodes described a kaleidoscopic drifting from perspective to perspective that he exploited for esthetic purposes: "When I am awake, I can look at a tree or a cat or a bird and see the air around it; sometimes it looks like water, so sometimes I paint water. Sometimes there is sound in the water so I paint sound. Sometimes the sound is silent like a fish swimming so I paint fish swimming" (quoted in Sass 1992, p. 136). "I can go on from dimension to dimension in thinking about something," said one patient of mine.[8] At times this may derive from a kind of "intellectual ambivalence"—what another patient called a "countervailing tendency" of the mind (pp. 463, 129).

Schizophrenic vacillation among frames of reference differs from that of concretistic organic patients, who do not as readily set up or move among alternative hypotheses, or show the same sense of ambivalence, conflict, hesitation, or self-awareness. It also differs from the distractibility of mania, in which attention moves rapidly but usually in a less radical, confusing, and self-paralyzing way. There is something fluid, spontaneous, and exuberant about the manic person's rapidly shifting attention, which typically moves among ideas existing within a single context—as in Foucault's "the incongruous." By contrast, the schizophrenic often seems aware of several possibilities, moving, or hesitating, among what seem to be alternative worlds or orientations.

[8] Schafer (1954) describes such "dramatic" shifts as "almost distinctively schizophrenic" (p. 107). M. Bleuler (1978) notes that "the world [the schizoid person] experiences undergoes a constant, renewed mental re-adaptation" (p. 435).

Conclusion

As we have seen, the disturbances of cognition in schizophrenia comprise a variety of abnormalities in the organization of attention, perception, thought, and language—including perplexity, dissemination, fragmentation, alienation, and slippage of various kinds. Especially common in schizophrenia—but not in mania, agitated depression, or most organic conditions—are various types of perspectival shift or drift that dissolve the sustained "perspectival abridgement" (the automatic blocking-out of alternative perspectives) that is necessary for practical action or clear communication, while often encouraging hyper-reflexive preoccupations of a hyper-abstract, quasi-philosophical, or overly literal kind (Sass 1992, chaps. 4, 5; Sass 2000; Sass & Parnas in press). The characteristic modes of schizophrenic attention include an overly idiosyncratic, goalless, and meandering stance (Bellack et al. 1990) devoid of an ongoing sense of the probable, as well as an overly analytic, detail-oriented, and intellectual orientation replacing one that is intuitive or holistic.[9] The person with schizophrenia often fails to adopt a practical, goal-oriented stance or attitude toward the world, or to constitute the associated pattern of meanings that make for a coherent and significant world. We have seen that the characteristic abnormalities of schizophrenic consciousness are not, or not primarily, a consequence of the loss of the basic ability to perceive forms, to focus, or to filter out the irrelevant. They involve, rather, the experience of context, of what Gurwitsch calls the thematic field.

[9] Zubin (1975) argues that what may seem to be indications of disturbed selective attention, maintenance of attention, and slowed reaction time may in fact be a function of "the goalless uniquely idiosyncratic life" or the unusual "culture" and past experience of the typical schizophrenic person (pp. 146, 160).

References

Angyal, A. 1964. Disturbances of thinking in schizophrenia. In J. S. Kasanin (Ed.), *Language and Thought in Schizophrenia.* New York: Norton, pp. 115-123.

Bellack, A.S.; Mueser, K.T.; Morrison, R.L.; Tierney, A.; et al 1990. Remediation of cognitive deficits in schizophrenia. *American Journal of Psychiatry*, *147*: 650-655.

Carr, V. J.; Dewis, S. A. M.; & Lewin, T. J. 1998. Preattentive visual search and perceptual grouping in schizophrenia. *Psychiatry Research*, *79*: 151-162.

Chey, J. & Holzman, P. S. 1997. Perceptual organization in schizophrenia: Utilization of the Gestalt principles. *Journal of Abnormal Psychology, 106*: 530-538.

Cohen, J. D.; Barch, D. M.; Carter, C.; Servan-Schreiber, D. 1999. Context-processing deficits in schizophrenia: Converging evidence from three theoretically motivated cognitive tasks. *Journal of Abnormal Psychology, 108*: 120-133.

Cohen, J. D.; Targ, E.; Servan-Schreiber, D.; Spiegel, D. 1992. The fabric of thought disorder: A cognitive neuroscience approach to disturbances in the processing of context. In D.J. Stein & J.E. Young (eds*), Cognitive Science and Clinical Disorders*, pp. 101-127. New York: Academic Press.

Conrad, K. 1958. *Die beginnende Schizophrenie*. G. Thieme.

Cutting, J. 1985. *The Psychology of Schizophrenia.* Edinburgh: Churchill Livingstone.

Derrida, J. 1981. *Dissemination*, transl. by B. Johnson. Chicago: University of Chicago Press.

Dworkin, R.H.; Oster, H.; Clark, S.C. and White, S.R. 1998. Affective expression and affective experience in schizophrenia. In M.F. Lenzenberger & R.H. Dworkin (Eds.), *Origins and Development of Schizophrenia.* Washington D.C.: American Psychological Association, pp. 385-424.

Foucault, M. 1994. *The Order of Things*. New York: Vintage Books.

Frith, C. D. 1979. Consciousness, information processing, and schizophrenia. *British Journal of Psychiatry, 134*: 225-235.

Frith, C.D. and Allen, H. 1988. Language disorders in schizophrenia and their implications for neuro-psychology. In P. Bebbington & P. McGuffin (Eds.), *Schizophrenia: the Major Issues*. Oxford: Heinemann, pp 172-186.

Frith, C.D.; Stevens, M.; Johnstone, E.C.; Owens, D.G.C.; Crow, T.J. 1983. Integration of schematic faces and other complex objects in schizophrenia. *Journal of Nervous and Mental Disease, 171:* 34-39.

Gray, J. A.; Feldon, J.; Rawlins, J. N.; Hemsley, D. R.; Smith, A. D. 1991. The neuropsychology of schizophrenia. *Behavioral and Brain Sciences*, *14*: 1-20.

Goodarzi, M. A.; Wykes, T.; Hemsley, D. R. 2000. Cerebral lateralization of global-local processing in people with schizotypy. *Schizophrenia Research*, *45*: 115-121.

Gurwitsch, A. 1964. *The Field of Consciousness*. Pittsburgh PA: Duquesne University Press.

—. 1966. Phenomenology of thematics and of the pure ego: Studies of the relation between Gestalt theory and phenomenology [Gurwitsch's thesis, originally published, 1929]. In A. Gurwitsch, *Studies in Phenomenology and Psychology*, Evanston IL: Northwestern Univ. Press, pp. 175-286.

—. 1985. *Marginal Consciousness*. Athens, Ohio: Ohio University Press.

Hamilton, M. (ed) 1976. *Fish's Schizophrenia, 2nd Edition*. Bristol UK: John Wright.

Heidegger, M. 1962. *Being and Time*, transl. by J. Macquarrie & E. Robinson. New York: Harper & Row.

Heinrichs, R. W. 2001. *In Search of Madness: Schizophrenia and Neuroscience*. New York: Oxford University Press.

Hemsley, D. R. 1976. Attention and information processing in schizophrenia, *British Journal of Social and Clinical Psychology, 15*: 199-209.

Holzman, P. S.; Shenton, M. E.; Solovay, M. R. 1986. Quality of thought disorder in differential diagnosis. *Schizophrenia Bulletin, 12*: 360-372.

Holzman, P. H. and Johnston, M.H. 1979. *Assessing SchizophrenicThinking*. San Francisco: Jossey-Bass.

Jones, S. H.; Hemsley, D. R.; Gray, J. A. 1991. Contextual effects on choice reaction time and accuracy in acute and chronic schizophrenics: Impairment in selective attention or in the influence of prior learning? *British Journal of Psychiatry, 159*: 415-421.

Knight, R.A. 1984. Converging models of cognitive deficit in schizophrenia. In W. D. Spaulding & J. K. Cole (eds), *Theories of Schizophrenia and Psychosis: Nebraska Symposium on Motivation 1983*, pp. 93-156. University of Nebraska Press.

Knight, R.A. and Silverstein, S. 1998. The role of cognitive psychology in guiding research on cognitive deficits in schizophrenia: A process-oriented approach. In M.F. Lenzenweger & R.H. Dworkin (Eds.), *Origins and Development of Shizophrenia*. Washington D.C.: American Psychological Association, pp. 247-295.

Kuperberg, G. R.; McGuire, P. K.; and David, A. S. 1998. Reduced sensitivity to linguistic context in schizophrenic thought disorder: Evidence from on-line monitoring for words in linguistically anomalous sentences." *Journal of Abnormal Psychology, 107*: 423-434.

Lecours, A.R. and Vanier-Clement, M. 1976. Schizophasia and jargonaphasia. *Brain and Language, 3*: 516-565.

Liddle, P. F.; Carpenter, W. T.; Crow, T. 1994. Syndromes of schizophrenia: Classic Literature. *British Journal of Psychiatry, 165*: 721-727.

Matussek, P. 1987. Studies in delusional perception. In J. Cutting & M. Shepherd (eds), *The Clinical Roots of the Schizophrenia Concept*, pp. 89-103. Cambridge UK: Cambridge University Press.

McGhie, A. & Chapman, J. 1961. Disorders of attention and perception in early schizophrenia. *British Journal of Medical Psychology, 34*: 103-116.

Merleau-Ponty, M. 1962. *The Phenomenology of Perception*, transl. by C. Smith. London: Routledge & Kegan Paul.

Peuser, G. 1987. Jargonaphasia and schizophasia. In R. Wodak and P. Van de Craen (eds). *Neurotic and Psychotic Language Behaviour.* Clevedon, Avon, UK: Multilingual Matters Ltd, pp. 216-248.

Phillips, W. A. and Silverstein, S. M. (in press). Convergence of biological and psychological perspectives on cognitive coordination in schizophrenia. *Behavioral and Brain Sciences.*

Place, E.J.S. and Gilmore, G.C. 1980. Perceptual organization in schizophrenia. *Journal of Abnormal Psychology, 89*: 409-418.

Polyakov, U.F. 1969. The experimental investigation of cognitive functioning in schizophrenia. In M. Cole & I. Maltzman (eds), *A Handbook of Contemporary Soviet Psychology*, pp. 370-386. New York: Basic Books.

Rapaport, D.; Gill, M.; Schafer, R. 1968. *Diagnostic Psychological Testing.* New York: International Universities Press.

Sass, L. 1992. *Madness and Modernism: Insanity in the Light of Modern Art, Literature, and Thought.* New York: Basic Books (Harvard University Press paperback 1994).

—. 2000. Schizophrenia, self-experience, and the so-called "negative symptoms" (Reflections on hyperreflexivity). In D. Zahavi (ed), *Exploring the Self: Philosophical and Psychopathological Perspectives on Self-Experience.* Amsterdam: John Benjamins Publ. Co., pp. 149-182.

Sass, L. and Parnas, J. (in press). Schizophrenia, consciousness, and the self. *Schizophrenia Bulletin.*

Schafer, R. 1954. *Psychoanalytic Interpretation in Rorschach Testing.* New York: Grune & Stratton.

Sechehaye, M. (ed) 1970. *Autobiography of a Schizophrenic Girl.* New York: New American Library.

Servan-Schreiber, D.; Cohen, J.D. & Steingard, S. 1996. Schizophrenic deficits in the processing of context. *Archives of General Psychiatry, 53*: 1105-1112.

Shakow, D. 1962. Segmental set: A theory of formal psychological deficit in schizophrenia. *Archives of General Psychiatry, 6*: 1-17.

Shenton, M.E.; Solovay, M.R.; & Holzman, P.S. 1987. Comparative studies of thought disorder: II. Schizoaffective disorder. *Archives of General Psychiatry, 44*: 21-30.

Silverstein, S. M.; Kovacs, I.; Corry, R.; Valone, C. 2000. Perceptual organization, the disorganization syndrome, and context processing in chronic schizophrenia. *Schizophrenia Research, 43*: 11-20.

Silverstein, S. M.; Matteson, S.; Knight, R. A. 1996. Reduced top-down influence in auditory perceptual organization in schizophrenia. *Journal of Abnormal Psychology, 105*: 663-667.

Silverstein, S. M.; Schenkel, L. S. 1997. Schizophrenia as a model of context-deficient cortical computation. *Behavioral and Brain Sciences, 20*: 696-697.

Spitzer, M. 1997. A cognitive neuroscience view of schizophrenic thought disorder. *Schizophrenia Bulletin, 23*: 29-50.

Störring, G. (1939/1987). Perplexity. In J. Cutting & M. Shepherd (eds.), *The Clinical Roots of the Schizophrenia Concept*. Cambridge, UK: Cambridge University Press, pp 79-82 (orig. published 1939).

Yoshimi, J. (this volume). Field theories of mind and brain.

Zubin, J. (1975). Problem of attention in schizophrenia. In M. L. Kietzman, S. Sutton, J. Zubin (eds), *Experimental Approaches to Psychopathology*. New York: Academic Press, pp. 139-166.

Chapter 4

Intentionality, Consciousness, and Intentional Relations: From Phenomenology to Cognitive Science[1]

John Barresi
Dalhousie University

Abstract

In this chapter I look closely at the intentionality of consciousness from a naturalistic perspective. I begin with a consideration of Gurwitsch's suggestive ideas about the role of acts of consciousness in constituting both the objects and the subjects of consciousness. I turn next to a discussion of how these ideas relate to my own empirical approach to intentional relations seen from a developmental perspective. This is followed by a discussion of some recent ideas in philosophical cognitive science on the intentionality of consciousness, both with respect to the objects and the subjects of consciousness. I show that these recent trends tend to naturalize intentionality and consciousness in directions compatible with the descriptive aspects of Gurwitsch's constitutive phenomenology.

Introduction

Cognitive science is currently undergoing a major transformation, one with an increased focus on the phenomenology and function of consciousness. It is no longer sufficient to provide a functional account of behavior understood in terms of cognitive and neural mechanisms. It is now also necessary to account for the experiences we undergo in the process of producing behavior and to give a naturalistic account of how these experiences might play a functional role in generating behavior. In this context, the descriptive phenomenology of consciousness of Aron Gurwitsch seems likely to supply an especially fruitful contribution to the study of consciousness. In this chapter I show how his views on the intentionality of consciousness, in particular, anticipate, in major respects, views recently developed by cognitive scientists, including myself.

[1] I wish to thank Lester Embree for organizing the research symposium on Gurwitsch's Relevancy for Cognitive Science and he as well as other participants in the symposium for their comments on earlier versions of this paper, both oral and written. I wish also to thank the SSHRC of Canada for providing research support and part of the funds necessary to attend the symposium.

L. Embree (ed.), Gurwitsch's Relevancy for Cognitive Science, 79-93.

Gurwitsch's (1964, 1966, 1984) descriptive analysis of acts of consciousness and how they relate to each other and to the notions of ego and object, seems, to me, to be just right. In the case of the objects of consciousness, Gurwitsch uses Husserl's notion of perceptual noema to provide a constitutive phenomenology of our intentional relations to real objects. His, and Husserl's, idea that different conscious acts can relate to a particular real object by some meaning or content that can track the identity of the object through multiple particular engagements, seems right to me. For instance, the Gestalt principle of good continuation would seem to be a rule by which we can keep track of particular objects in motion, even though, sometimes, there may be a switch in objects, so that an object is the "same" only in the content intended by the conscious act, not by the object actually referred to—perceptually—in the act. The identity of the "intended" object must carry across particular perceptual acts by some form of logical or semantic rule of identity of this sort.

I would not want to say that this constructive approach to content is the only way that consciousness relates to objects. As a psychologist, rather than a phenomenologist, I must assume the reality of the object, and of the constraints and opportunities it affords for further interaction and identification. And Gurwitsch's view of psychology as a naturalistic science as well as a phenomenological science to some extent bridges this gap between his phenomenological psychology and my own view (see, Embree, 2003, for a discussion of Gurwitsch on these two psychologies). Whereas in his phenomenological psychology, Gurwitsch considers the intentionality of consciousness alone, while treating the world represented in conscious acts, including our own bodies and minds, as objects taken as real only as intended by these acts, when he considers psychology as a natural science, he admits the psychophysical being as the object of investigation, not the phenomenological being, or stream of consciousness, by itself. Still, it is not clear how far he would be willing to go in relating his notion of natural scientific to phenomenological psychology. While I agree with his pronouncements that any adequate scientific psychology must take into account discoveries in pure phenomenology and its mundane interpretation as phenomenological psychology, I am not sure Gurwitsch would accept a need also to go in the opposite direction. To what extent would he be willing to reinterpret the findings of adult phenomenological psychology based on discoveries of how the child or animal mind works, or on laboratory studies of consciousness (e.g., Christie & Barresi, 2002)? While he talks often about reflecting on the mental states of others, including animals, children, and those with mental disorders within the context of developing a phenomenological psychology, I am not sure how far he would accept such findings as having implications on whether or how phenomenological psychology, itself, could be incorporated within a natural scientific psychology to form a unified naturalistic science of organisms and persons. Gurwitsch's view would seem to be that phenomenological psychology must always have precedence over naturalistic scientific psychology, thus forever limiting how the former might be integrated within the latter.

Thus it may be that a pure constitutive phenomenology, and its mundane interpretation as phenomenological psychology, in Gurwitsch's sense, might not go quite far enough in a naturalistic direction for naturalists like myself, who are interested in developing a science of persons that includes an evolutionary and developmental as well as a cultural perspective on the notion of person (Barresi, 1999). Nevertheless, I believe that Gurwitsch is still surely right about the importance of Gestalt-like constitutive principles for how in human development we first come to know the external objects that we find to exist in conscious intentional acts. Moreover, what particularly intrigues me about Gurwitsch's views is that in addition to his attempt to provide a phenomenological account for identity in the external object of intentional relations, he also attempts to do the same for the subject of intentional relations. I believe that his non-egological view of intentional acts of consciousness (Gurwitsch, 1966), which anticipated Sartre's similar rejection of the later-Husserlian notion of a pure ego as a central and identical transcendental source of all conscious activity, connects quite closely to some very recent thinking on the same topic, including in my own work.

Gurwitsch writes of the ego:

> What is meant by the ego is nothing apart from this united complex. It derives its unity and its coherence from the very acts that enter into it and constitute it; and it is nothing other than the organized totality of these acts. Hence when the subject, reflecting upon the act he experiences, ascertains that this act is his, this only means that the act in question is a part of that complex and has its place within this united and organized whole. In this theory there is obviously no place for a center or a pole of conscious life from which the acts might issue or emerge. (Gurwitsch, 1966, p. 288)

And on the issue of knowledge of my own versus another person's ego, he writes:

> My ego and my psychic facts, in contradistinction to the conscious acts, are then no longer my exclusive property, for they are accessible to other people, whereas my consciousness is not so; it is and remains closed and impenetrable for everyone except myself. The problem of the comprehension of other persons' minds is thus simplified and must be raised in quite new terms. The condition, however, of this simplification is the non-egological conception of consciousness. (Gurwitsch, 1966, p. 297)

Gurwitsch's sensitivity here to how the empirical ego can only come into view through reflective consciousness and that the ego or self perceived—or, rather, constituted—in this way, is much the same for self and other, is strikingly similar to a view of the understanding of self and other that I have been developing in separate collaborations with Chris Moore and with Ray Martin. In order to show this similarity, as well as to show other similarities—and differences—in our views, I will describe next the theory we have been developing, which Chris Moore and I now call Intentional Relations Theory (or IRT).

I. Intentional Relations Theory

As originally presented in our 1996 *Behavioral and Brain Sciences* target article, IRT was developed to provide a phylogenetic as well as an ontogenetic account of how organisms come to understand their own minds as well as the minds of others. In particular, we were interested in explaining how organisms come to interpret the intentional activity of self and other within a common conceptual system. The puzzle here is that we experience our own activity from a first-person perspective, and the comparable activity of others from a third-person perspective, and that these two perspectives have little in common. Yet we, as adult human agents, are able to recognize the intended objects of actions of others, the object-directed emotions and motivations of others, and the object-directed epistemic states of others, in pretty much the same terms as we understand these intentional activities when we ourselves are engaged in them. How is this possible?

A traditional answer to this question of how we understand the intentional relations of others is that we model their objective activities, which we can only know directly as behavior, by imagining from a first-person perspective, what we would experience when doing what they are doing. However, this knowledge of other minds by "analogy" or by what is now called "simulation" cannot be the way that we initially learned to understand others. For this simulation view does not take into account the problem of how we come to know our own minds or egos, which as Gurwitsch realized, is not immediately transparent in first-person conscious activity.[2] How do we, then, come to recognize that the behavior of others is the same as our own comparable behavior, and come to realize that they have first-person conscious and intentional properties that go along with their behavior, but which are not directly perceivable in the behavior? The problem is that even in our own case, we have no a priori basis for distinguishing between our subjective first-person intentional states and their objective properties

[2] See Gallagher's chapter in this volume for a more comprehensive Gurwitschean critique of the simulation view.

involving our bodies. So how do we acquire this knowledge that is needed to perform the simulation?

According to the theory that Moore and I presented in the BBS article, the understanding of intentional relations of self and other depends on an initial understanding of shared or matched intentional relations of self and other. By intentional relations, we mean, in our theory, the psychophysical relations that organisms have with objects actually in their environment as well as with intentional objects. Thus our starting point is a naturalistic understanding of intentionality. By shared intentional relations we mean situations when two or more organisms, typically self and other, are engaged concurrently in the same kind of intentional relation to the same or similar objects, and when an individual is at the same time aware of this joint activity of self and other. Such shared intentional relations can come about through imitation involving the same or similar objects, contagious emotional sharing sometimes involving some particular object or person, joint attention to some object or event, or some other form of joint activity involving objects. In such situations, through what Moore and I call the intentional schema, we hypothesize that humans, and, possibly, some other organisms, are able to link together into a single concept or schema, the third-person properties of the intentional relation as we perceive it in the activity of the other, with their first-person properties as experienced in self when also engaged in this same activity. This allows us to understand intentional activities both from a first- and a third-person perspective. Later, through the use of this schema in memory, we can come to recognize intentional relations of the same sorts previously jointly experienced with others, in situations not involving shared or matched activities. Having integrated the first- and third-person properties of these intentional relations in these previous joint activities, we can now recognize them from their first-person properties taken alone in our own case or from their third-person properties taken alone in the case of others.

It may help here to give an example. When we see someone turn to look at an object and show interest in it, we tend, almost automatically, to turn in the same direction to look at the object. From a phylogenetic point of view, the ability of social organisms to respond in this way to the attentional activity of others is quite functional. Nevertheless, this activity, as such, does not require any understanding, from a first-person point of view, of the intentional relation that the other is engaged in. The third-person information is enough to elicit the appropriate functional response. Indeed, even a nine-month-old human infant can engage in such matching activity, without also being aware of the similarity of first- and third-person properties of self and other in this activity.

But not long after nine months the human infant shows evidence of being aware of the internal or first-person aspect of the other's point-of-view in such intentional states. For instance, at roughly this age, the infant sometimes will point to an object, apparently in order to draw the attention of an adult to the same object, while appreciating that the adult also turns to the object. And a short while later, the infant, who sees the sadness of another child, may respond with sympathy to the other's need, whereas earlier on the infant would focus only on her own sadness, which was acquired through contagion from the other's crying.

What Moore and I suggest in our theoretical account of these phenomena is that the infant's reflective awareness of intentional relations of self and other initially develops out of events involving shared activity with others, where the infant experiences her own first-person aspect of an intentional relation, while, at the same time, observing the third-person aspect of the same or similar intentional relation of the other person. Through the use of an intentional schema, these two aspects are integrated into a single representation of that intentional relation, which contains both first- and third-person aspects. At this early phase, the infant is not aware of the distinction between the intentional activities of self and other, so the representation that the infant has of the intentional relation, is fundamentally intersubjective and spans across self and other. Yet, because the representation allows for first- and third-person aspects, this representation of the shared intentional relation lays the foundation for future individualistic understanding of intentional relations, which develops in humans toward the end of the second year.

According to our account, what makes possible the individualistic understanding of self and other intentional relations at the end of the second year is the development of the infant's imagination, so that the infant can hold in mind an imagined aspect of an intentional relation of a particular person at the same time as holding in mind a perceptual aspect of the same intentional relation. This applies both to understanding one's own intentional relations and to understanding the intentional relations of others, though the perceived and imagined aspects differ in the two cases. In the case of others, the perceived aspect is the overt behavior of the other of which one is directly aware—e.g. crying—and the imagined aspect is the inner experience that one imagines the other to be in when crying. In the case of self, the reverse occurs in that the perceived aspect is the inner feeling of emotion and the imagined is the outward expression of that emotion.

Let's go back to the instance of sympathy to see the developmental process involved here. Whereas the one-year-old infant cries when another infant cries and tends only to console herself, gradually the infant comes to recognize the similarity in states between self and other, matching, in effect, the first-person aspect of her own crying, with the third person aspect of the other's crying. Eventually, she realizes that it is the other infant who is originally the sad one, at which time her own sadness is understood as originating in the internal state of the other. Her sympathetic response to the other grows during this period, from a simple kind of sympathetic sharing, say of her doll, with the other, to a realization that what the other may find more soothing is her own doll, or her own mother, not that of the infant herself. This more advanced sympathetic response requires imagination, as it requires distinguishing the infant's own intentional relations from those of the other infant. However, on our account, what provides the content of what one imagines here is one's own previous experiences being in a similar state, which can be recognized as such only from having shared in the past similar states with others.

At the end of the second year, or shortly thereafter, the infant also develops the ability to be embarrassed about her own appearance to the other. We hypothesize that what happens here is that the infant can now imagine how she looks from a third-person perspective, when she has direct first-person experience of the activity she finds embarrassing. Again, it is from previous experience, for instance, from observing others in similar circumstances and sharing in their experience, which provides the basis for such self-reflected states of embarrassment. In this case, the intentional schema provides the imagined third-person component, which became initially associated with the first-person aspect, when one previously observed and also shared in the other's similar experience.

In all of these individualistic cases, whether of self or of other, we propose that the infant has a generalized representation of an individual agent in an intentional relation to an object or situation, and we propose that this representation includes both first- and third-person properties. So, it is a representation of the intentional state of what Gurwitsch might call the empirical ego, the ego of immediate agency, as well as of dispositional properties, which link the agent across time. And, as in Gurwitsch's account, we see that this perspective on the empirical ego can be adopted both by the person herself, and by another person; but that such knowledge involves reflection on simple intentional relations involving an agent and object. On this view, the ego is not originally, and not typically, contained within the first-person intentional consciousness of an agent, which focuses on the object rather than on the agent of the intentional relation, but only appears within reflective consciousness, when third-person objective properties of the agent's existence in the world are added to these first-person subjective properties.

It is important to note, however, that in our BBS account of the two-year-old's understanding of intentional relations, Moore and I did not suppose that the child understands the representational nature of mental states, and in this respect, we may differ to some degree with Gurwitsch's account,[3] which tends to focus more on the ego as a mental being, than on the person as a psychophysical, intentional being. Our claim is that what the child understands at this time is that she and other people have an inner and outer aspect with respect to their relations to objects or goals. She places herself and others in the world, and conceives of herself and others as bodied agents, not mental agents with bodies. In order to reach the level of understanding of self and other as mental agents, we propose,

---

[3] This is the case at least with respect to the account presented in his paper, "A Non-Egological Concept of Consciousness," Gurwitsch, 1966, pp. 287-300.

with other cognitive scientists, that the child needs to know the difference between how an individual represents reality and the reality that is represented, where by reality here I mean nothing more than the commonsense everyday world of experience.

Evidence from a number of experimental tasks indicates that this more advanced conception of mental agent first appears around the age of four. At this time, the child becomes able to know that another person's representation of reality may be in error, and that her own representation of an object may not be the same as another person's representation of the same object. In experiments testing the child's understanding of false belief, the child sees a puppet put an object in one location, then leave the room. While this puppet is gone, another one moves the object to a new location. When the first puppet returns, the child is asked where it will look for the object. Three-year-olds say that the puppet will look in the new location. After all, that is where the object is. But the four-year-old will predict correctly that the puppet will look where it put the object, thereby showing that she understands that her own representation, or understanding, of the location of the object is not the same as the representation that the puppet has of the object's location. Thus, at this time, the child acquires the ability to understand that intentional relations may occur where the object of those relations, the content of those relations, may not exist as real objects in the world, but only exist in the individual mind. In our account, the child achieves this capacity only when she can imagine both the first- and the third-person component of an intentional relation, and can compare it to reality, at least as she supposes it to be.

It is also interesting to note that it is at this same age-four transition that the child begins to develop a notion of a temporally extended self—a self that transforms through time in a causally and intentionally connected fashion. The acquisition of this notion of an extended self seems tied to the child's ability to become aware of her own current mental states as mental as well as to be able to think back to prior mental states and to anticipate future mental states. One source of evidence indicating this transformation, with a focus on past mental states, comes from a task where the child is asked what she thinks is in a box, with pictures of a candy—Smarties—on the cover. After the child says candy or Smarties, the box is opened and it is discovered to have pencils in it. Shortly thereafter the child is asked to say what someone else, who had not seen the box opened, would think was in the box, and also asked what she thought was in the box before opening it. For three-year-olds, both questions are typically answered with "pencils," but for four-year-olds, who now understand the possibility of false beliefs, the answer to both questions is "candy." This result suggests that being able to understand the representational, or intentional, nature of mental states is necessary to answer these questions. Other evidence suggests that distinguishing between present and future mental states of self develops at this same time. Taken together, these and other results seem to support the notion, suggested by Gurwitsch, that the ability to reflect on one's past (or future) mental states (i.e., one's stream of conscious intentional acts in his formulation), is necessary in order to conceive of one's self as an empirical, mental ego extended in time.

From a developmental point of view there is still more to the story. Other major transformations in our self or ego-concepts occur. In particular, it is not until late adolescence or early adulthood that at least some people go through an existential and/or identity crisis as a result of which they are confronted with their existential freedom and come to view themselves as agents who can construct themselves or their identities in the process of living, as well as generate a narrative, which both describes and justifies who they are. Although not part of his own formulation, even this later development provides evidence to support Gurwitsch's non-egological orientation to the subject of intentional consciousness and to the role of a phenomenologically constituted empirical ego in mental activities.

II. Intentional Relations and Consciousness: Thick and Thin; Narrow and Wide

At this point I want to turn to the phenomenological problem of consciousness and intentional relations. So far my approach to intentional relations has been from the point of view of the natural everyday world of folk psychology. In other words, the focus has been on how we come to understand intentional activities of self and other in our everyday relations with each other and the world. I now want to focus attention on the structure of consciousness involved in intentional relations and to provide a kind of constitutive phenomenology of this structure. You will have to bear with me as a psychologist entering this domain, because my leanings will be on the psychological rather than on philosophical implications of this analysis. Yet I hope you will find it a fruitful endeavor, and a possible meeting ground of common interests.

As the present book and the conference upon which it is based is a sign, there has been a recent intense interest in the investigation of consciousness from a multi-disciplinary perspective. Another such recent conference, or rather NEH institute, was held at Santa Cruz last summer (2002), where a number of interesting ideas were discussed. Although I did not attend the institute, I have read some of the papers discussed there and have found them to provide some ideas worth pursuing here. The purpose of the summer institute was to look into the relationship between consciousness and intentionality. Based on the readings for the institute, I found two notions particularly interesting to the psychology and phenomenology of intentional relations and consciousness and of particular relevance to my understanding of Gurwitsch's contributions that I have been discussing in this chapter.

The first idea is a distinction that is currently being made between "narrow" and "wide" interpretations of the intentionality of consciousness. This distinction has been clearly made in a recent paper by Terence Horgan and John Tienson (2002). The main reason they have for making the distinction is to deal with how intentionality is to be understood in our consciousness of objects when viewed naturalistically. But it also relates to the classic problem of the phenomenological reduction; and, as I understand it, the distinction might provide a way for integrating a Gurwitschean approach with its emphasis on the objects as constituted in phenomenology itself, with a more traditional Husserlian approach where the object, itself, is taken to be an "x" that stands outside of the reduction.[4] What the narrow versus wide distinction focuses on is whether the phenomenological "intention" should be understood as directed at the object as constituted in the phenomenology or at the actual object represented by that phenomenology. In the narrow view, all that matters here is the phenomenology itself and what the object must be like if it is to fulfill the intention. But no attempt is made to go beyond the constitutive structure of this intention, which includes the object as it appears, though not the object as it is in itself. By contrast, when one takes a wide perspective on the intentionality of conscious phenomenology, the interest here is on the success or failure of the intention to determine the actual object and on the causal processes involved in specifying what the "true" intentional object is.

For those not previously exposed to this distinction, it might help to see a typical example that illustrates it. Consider Donald Davidson's Swampman, who replaces the true person, recently vaporized by lightning. The person that Swampman replaces is supposed to have a phenomenology identical with Swampman. However, whereas the original person truly perceived and remembered objects; at least initially, if one takes an externalist or what here is called the "wide" view of intentional consciousness, Swampman can neither remember nor perceive anything, because he lacks the causal history of interactions necessary to truly remember or perceive. For instance, when he "remembers" being in a certain situation in the past, he neither was truly in that situation, nor was he the person he thinks he was in that situation. Furthermore, even if he "sees" a tree before him "as a tree" he fails in his categorization, because he has not had the causal history to warrant using the concept "tree" to that object apparently before him. Taking the wide or externalist view, Swampman fails to have intentional consciousness of any objects because he lacks the representational wherewithal that comes about only through appropriate causal interactions that constitute a phenomenology as representational of objects. But if we allow for either a narrow or phenomenologically reduced interpretation of intentional consciousness, Swampman can be said to have the same intentional consciousness as the original person. After all, phenome-

[4] See David Smith's chapter for an alternative integration of Husserl and Gurwitsch from a phenomenological perspective.

nologically he does seem to perceive and remember in the same way as the original person. By making the distinction between narrow and wide, we are thus able to attribute a consciousness to Swampman, even if, ultimately, or at least in some circumstances, we want to view the content of consciousness in a wide sense of content, in order to deal with actual objects being represented.

There are two things to note about this distinction between narrow and wide content that one can readily exploit. The first is that if one has the right kind of history and an agreeable world, then wide content will, de facto, follow narrow content rather closely. The second is that because of this strong relation, one can readily imagine a psychology of development that exploits the difference between narrow and wide content, where narrow content would represent first-person intentional consciousness and wide content would represent what we, as psychologists and phenomenologists, take a person's intentional relations actually to be.

One can also see how Gurwitschean themes might be pursued using this distinction. For instance, even Swampman might be said to have intuitive perceptual noemata. When he sees the tree, surely he has a direct noematic intuition of the tree that is part of his "perception" of the tree "as a tree," however deceived he might be with regard to his prior acquaintance with trees. But one can go further with the distinction between narrow and wide, used in this manner, in order to begin to discuss the infant's development of perceptual knowledge of the world of objects. An infant, with little prior experience with the world, will have a much undeveloped range of perceptual content, when viewed narrowly in immediate first-person intentional consciousness, though in a wide sense, the infant may be in intentional relations with a diverse range of objects. But because the infant cannot provide discriminative responses to these objects, some are treated as identical to each other, even though, in a sense, his perceptual noemata, because of their intuitive richness, if utilized within a wider semantic set of noemata, might discriminate each of the objects from each other and track each of them through time. The developmental course of the acquisition of such discriminative semantic noemata, by which the child comes to know his perceptual world, then becomes an issue worthy of investigation. But rather than follow this topic further, let me turn now to the other side of the coin, from intentional content or objects to what Ray Martin and I (Martin & Barresi, 2002) have called "intentional subjects."

A major concern of any non-egological psychology is to explain the development of our sense of personal identity through time. I have already discussed certain landmarks in the infant's development of a sense of self that she applies to self and other. But here I want to pursue the issue further, and look more closely at the phenomenological problem that is being faced whenever we connect our self across time and attribute to ourselves any kind constancy or continuity in our essence.

In order to get into this issue, I want to focus on recent writings by Galen Strawson (1997, 1999, 2002a, 2002b), including some papers recently written for the Santa Cruz summer institute. Strawson has been working on ideas about the self—taken as the subject of consciousness—that he now recognizes as having a ground in William James's discussion of self in consciousness. Like James, Strawson believes that there can be no "pure ego" that needs to be postulated as the psychological subject of thoughts (i.e., acts of consciousness), and like James, he holds that any connections made between the subjects of distinctly different thoughts must occur through some act by which the subject of any particular thought appropriates other thoughts as its own. Thus, any apparent identity between the subject of current experience (i.e., thought) and any past (or future) experience does not involve any essential identity in its subjects, only psychologically constructed identity. I hope that it is immediately obvious the Gurwitschean non-egological and phenomenologically constitutive nature of any process of self-construction of this type.

In earlier work (e.g., Strawson, 1997), Strawson was less aware of his relationship to James, and he also focused more on—what might be called—the Cartesian insight that the self, immediately given in an experience of a reflective type, necessarily exists as a unity, whether or not that self extends in any way into the past or future. And, at the time, he argued that selves of this type, if not typical of all people, were at the very least a possible mode of individual experience, one that did not require such a local or minimal self to appropriate past or future selves even of the same body. He thought of these local selves as a string of pearls associated with a continuously living human being. As pointed out by Arvidson (2000, p. 24), this metaphor of pearls on a string relates in an interesting fashion to a similar metaphor used by Gurwitsch in describing the stream of consciousness.[5]

In Strawson's most recent work for the NEH summer institute (Strawson, 2002 a, 2002 b; see also Strawson, 1999), he pushes this theoretical framework further, based on a phenomenological investigation of intentional consciousness, as well as some empirical considerations, such as how the brain might operate so as to constantly renew or create what James called "pulses of consciousness." At this point, Strawson wants to hold that all conscious acts involve a subjective pole and an objective pole that together form an "experience" *of* an "object" or "content" *for* a "subject." But he goes on from this phenomenological insight to the suggestion that it is fruitful to use a distinction between a "thick" and a "thin" notion of subject. Under the thick, or normal, everyday, notion, a subject can be viewed as extended in time and as being the bearer of multiple conscious intentional acts, all of which belong together as being acts by a single organism,

---

[5] Gurwitsch, 1966, p. 312-3; the metaphor may itself derive indirectly from James on this issue, see, e.g., 1966, p. 291-2.

or some part of that organism with its own continuous—but also partially discontinuous—consciousness. By contrast, he wants to argue for an empirically and philosophically useful concept of a thin subject. Indeed, in this recent conception, the subject is taken to be so thin that for every experience, treated as a unity of subject-experience-content, there is a subject and content, distinct from those of other experiences, no matter how similar in structure or close in time they are to these other experiences. Based on empirical evidence, Strawson thinks of such subject-experience-content units as existing for no more than 3 seconds.

In his further analysis Strawson takes these subject-experience-content units as the ontologically basic units of experience or consciousness. He thus agrees, in effect, with Gurwitsch and other phenomenologists on their notion of acts of consciousness. However, in this case, the emphasis is on a naturalized version of this notion. Strawson continues his analysis in order to work out what can be said further on an ontological level about the subject-experience-object correlation in such units of experience. His final conclusion is that, ontologically, there is a perfect identity among the three aspects of an experience or act of consciousness. So for every experience, from an ontological point of view, the "subject" of the experience is identical with the "object" of the experience and both are identical to the "experience" itself. He reaches this conclusion because, for such a thin subject, there is a perfect one-to-one correlation between subject and object—for every object there is one and only one subject pole, and for every subject, there is one and only one object pole, and together with their intentional relation, they form the experience. Or as he puts it in James' language, there is only the single isolated thought, which contains within it, when narrowly construed, both the thinker and object of thought. Such a thought or act of consciousness, when naturalistically construed, is thus the ontological primitive of all consciousness.

Although I have some reservations about Strawson's final conclusion that subject equals object because of their perfect correlation in an experience, I do find the enterprise intriguing, as it pushes to the limit what we can say about intentional relations viewed as single acts of consciousness. From a naturalistic perspective, what this implies is that every act or pulse of consciousness, when narrowly and thinly construed, has, not only an intentional object for some subject, but that the subject herself, is only an intentional subject, there being no "actual" subject separate from the conscious act. Such acts can then be said to organize themselves, either in terms of their objects, or their subjects, or both their subjects and objects, where all of these must be construed as intentional, not actual. Furthermore, from a naturalistic perspective such "organization" of acts of consciousness presumably occurs as the natural organization of meaning in the unifying activity of the brain. In other words, what we call unity of consciousness *just is* this unifying activity of the brain in pulses of consciousness, which contain within them intentional subjects, objects, and their relations, and which, when widely and thickly construed, gives us the natural world of agents, objects and intentional relations between them.

There is a sense in which Strawson's theoretical framework here is the naturalizing limit of the Gurwitschean analysis of acts of consciousness. Both Gurwitsch and Strawson emphasize the noetico-noematic correlation in acts of consciousness, both view the object as intentional content that when narrowly construed can carry over from one act to another only intentionally, not actually, and both, when focusing on the subject of particular acts of consciousness, see the subject non-egologically, where their connection to other subjects of consciousness must be made through their intentional similarity to other subjects, and because they are organizationally related to them. Both Gurwitsch and Strawson appreciate William James as an ancestor to their approach in trying to understand the organization of the field of consciousness into objects and subjects, though Gurwitsch speaks less of the subject aspect of James's theory and Strawson speaks less of the object aspect of James's theory. Where they differ is that Gurwitsch, as a phenomenologist rather than a philosophically oriented cognitive scientist, stresses the importance of acts of consciousness as the foundation of all knowledge. Strawson does not really disagree with Gurwitsch on this point, as can be seen in some of his work on mental realism (Strawson, 1994), but stresses, instead, the empirical usefulness of his analysis of acts of consciousness. But as we saw at the beginning of this chapter, neither is Gurwitsch alienated from naturalism. Nevertheless, it is doubtful that Gurwitsch would go as far as Strawson in naturalizing consciousness. For Gurwitsch, the analysis of the intentionality of consciousness has to serve as the foundation for all other investigations, and this does not seem consistent with a naturalistic assumption that makes consciousness a mere activity of brains.

In conclusion, I would like to suggest that the approaches that I have surveyed in this chapter are naturalistic extensions of Gurwitsch's ideas on the intentionality of consciousness, his concern to understand how objects can be intentional, and how subjects or egos are also intentional, though he may not have used the term in this context. Thus, although, he, himself, might feel that his constitutive analysis of phenomenology has a good bit more to offer philosophers than merely a reduction to naturalistic psychology or cognitive science, I believe that Gurwitsch would be sympathetic with our attempts in this chapter, and, more generally, in this book, to move in this direction.

References

Arvidson, P.S. 2000. Transformations in consciousness: Continuity, the self, and marginal consciousness. *Journal of Consciousness Studies*,7, 3, 3-26.

Barresi, J. 1999. On becoming a person. *Philosophical Psychology*, 12, 79-98.

Barresi, J., and Moore, C. (1996) Intentional relations and social understanding. *Behavioral and Brain Sciences*, 9:107-21.

Christie, J. and J. Barresi, J. 2002. Consciousness and information processing: A reply to Durgin. *Consciousness & Cognition: An International Journal*, II, 372-4.

Embree, L. 2003. Aron Gurwitsch's theory of cultural-scientific phenomenological psychology. *Husserl Studies*, 19, 43-70.

Gurwitsch, A. 1964. *The Field of Consciousness*. Pittsburgh, PA: Duquesne University Press.

—. 1966. *Studies in Phenomenology and Psychology*. Evanston: Northwestern University Press.

—. 1984. *Marginal Consciousness*, ed. Lester Embree. Athens, OH: Ohio University Press.

—. 1966. Husserl's theory of the intentionality of consciousness, in H. Dreyfus & H. Hall (eds.) *Husserl, Intentionality and Cognitive Science*. Cambridge, MA: MIT Press.

Horgan, T., and J. Tienson, J. 2002. The intentionality of phenomenology and the phenomenology of intentionality, in D. Chalmers (ed.), *Philosophy of Mind: Classical and Contemporary Readings*. New York: Oxford University Press.

Martin, R. and J. Barresi. 2002. Personal identity & what matters in survival: An historical overview, in R. Martin & J. Barresi (eds.), *Personal Identity*. Oxford: Blackwells.

Strawson, G. 1994. *Mental Reality*. Cambridge, MA: MIT Press.

—. 1997. The self, *Journal of Consciousness Studies*, 4: 405-428.

—. 1999. The self and the sesmet. *Journal of Consciousness Studies*, 6: 99-135.

—. 2002a. Consciousness and Intentionality (and Terminology) (draft 28 June) for NEH Summer Institute on Consciousness and Intentionality, UC, Santa Cruz (June 24-August 2)

—. 2002b. What is the relation between an experience, the subject of the experience, and the content of the experience? (Unfinished draft for NEH Summer Institute on Consciousness and Intentionality, Santa Cruz (June 24-August 2) eventually to be published in *Philosophical Issues*).

Chapter 5

The Experience of the Present Moment

Robert Piłat
Institute of Philosophy and Sociology, Polish Academy of Science, Warsaw, Poland

Abstract

In this essay I discuss an aspect of time awareness: the experience of the present moment. In "William James's Theory of the 'Transitive Parts' of the Stream of Consciousness," Aron Gurwitsch took up the problem of "the intrinsic temporality of an enduring act." Some of his remarks shed light on the nature of the experience of present moment. His theory of the field of consciousness provides grounds for the following suggestion: The experience of the present moment is built upon a specific structure of the presented content, namely, a double representation of each content of conscious experience: as a central part of the field and as a marginal part. The transition from the former to the latter is accomplished through a decomposing of the first presentation and re-building it in another part of the field. Accordingly, the experienced temporal flow is also a "movement" through the field. It presupposes a field structure. I find supporting insights for my view in Husserl's manuscript on temporization (C 6). I also refer to research into neural mechanisms of time awareness and to some recent theories of mental representation.

I. Intensional Contexts Containing the Term "Now" or its Synonyms

The radically naturalistic account of the experience of the present moment boils down to saying that events occurring in the experiencing subject and events in its environment take place in the same moment of physical time. An alternative account of the experience of the present moment is a cognitive construction whose realization within given subjects is loosely connected to the moment of physical time that the experience refers to. The task I am pursuing in this paper is to reconcile these two standpoints by combining cognitive and phenomenological approaches. Before attempting this, I will give reasons for going beyond the purely semantic analysis of sentences in which the experience of the present moment is reported.

The sentences describing experiences of the present moment, i.e., those containing the word "now" or its synonyms differ essentially from their paraphrases

L. Embree (ed.), Gurwitsch's Relevancy for Cognitive Science, 95-109.

containing the expression "in the moment." The former imply intensional contexts and the latter extensional contexts. According to the predominating semantics of intensional contexts, i.e., the semantics of possible worlds, the meaning of a sentence containing the word "now" is interpreted as a set of possible states of the world (speaking more generally: possible worlds) in which the sentence is true. Therefore, the semantic account of the experience of the present moment amounts to posing the question: What does it mean to say "I am experiencing x now?" In order to answer this question, we need ways of identifying an appropriate set of possible worlds. Were it accomplished, the experience of the present moment could be reduced to a certain state of the world and a relation between the sentence and the state.

It is doubtful, however, if the experience of the present moment can be accounted for in terms of the meaning of expressions that contains the phrase "in the present moment" or its synonyms. There are three reasons for this reservation:

A) Neglected intentionality. A purely semantic approach ignores the intentional character of the experience. The experience of the present moment does not imply that we utter or think sentences that refer to the states of the world contemporary with these utterances or thoughts (for such reference can be created by sentences uttered or thought later); rather, it implies that these words express a content of experience that orients itself intentionally towards the present state of the world. There are many ways of using the word "now" in which it maintains its meaning while its intentionality has nothing to do with the present experience of the speaker; for instance, in reported speech ("He says, he is in Paris now"), in sentences referring to other people ("Now he is visiting Paris"), or in sentences about other people stated in probabilistic mode ("He is probably visiting Paris now"). All these sentences denote the states of the world contemporary with acts of uttering these sentences—but they connote different intentions.

B) Non-tractable causality. For intentional contexts it is crucial that the state of the world contemporary with the utterance be causally connected with the utterance or thought. However, it is difficult to imagine how one can trace the causal links connecting:

(1) what happens in the present moment both outside and inside the organism of the experiencing subject; and

(2) the mental representation of the state of the world underlying the utterance containing "now" or its synonyms;

The state of organism in the given moment is a result of influences both of the external world and the former state of this organism. These two causal determinations have different temporal structures; organic processes (internal causality) impose a different measure of physical time than the external physical processes that are contemporary to the former. If we ever were to represent what happens outside and inside of the organism in terms of functions with a time variable, we would very likely end up with two sets of incompatible functions.

It seems plausible to see the sensory apparatus as a "spot" where the two hypothetical sets of functions get reconciled. This is accomplished by *standardization*, usually referred to as information. The emergence of information that happens on the basis of a complex network of internal and external causal relations takes place by means of numerous filters and auxiliary constructions that help to distill information from a causal nexus. The filtering functions are realized by neural networks and as such they need time. Therefore, the resulting state of the organism should be referred not to momentary events but rather to temporal phases of the world. Consequently, the very correspondence between the actual state of the organism and the actual state of the world appear loose. The causality that is responsible for the present states of organisms cannot be traced back to the well defined events or time slices of the world.

C) Model dependence. I might be tempting to say that sensuous creatures just have some sort of perception of the present moment—something similar to the perceptions of trees or buildings. Should this assumption be true, then the necessary and sufficient conditions of truthfulness of sentences like "X happens now" would be the same as time related conditions of identification of X by subject. The description of those conditions would not contain the word "now" or its synonyms. The sentence "A is happening now" would mean exactly the same as the sentence "A happens."

Such solution is counterintuitive though. Not only does it deprive expressions like "now" of any meaning but also ignores grammatical differences, which do reflect important difference in assumed models. Let's consider a simple example: "X happens" may refer to a "perforated" process like the changes of stock rates at the stock exchange that closes every evening and opens again next morning. When in the certain period of time the rates go down, one can say "Stock rates are going down" in every moment of this period, regardless the hour, although, literally speaking, every evening, after the closing the stock exchange they stop falling. This is not just a liberal approach to word usage. We use two different models for the process in question. On one model, there are acts of market players that influence the rate's change. This model divides time into portions and allows for breaks. On the other model, there is just a time continuum into which all economic value including the stock rates are mapped. This is a model we need to describe our economic reality.

Intentionality, causality, and model dependence are three features of the experience of the present moment that force a search beyond the semantic framework. In the remainder of this chapter I shall consider the cognitive point of view and dwell on the phenomenological one.

II. Mental Representation as the Basis of the Experience of the Present Moment

The causal connection between the actual state of the world and the state of the organism has a peculiar characteristic even at the purely physical level. There must, namely, be at least two states of organism, the internal cause and the internal effect, to correspond with a singular state of the world. States of organism result from one another and only as such are they connected to the state of the surrounding world. An object that reacts to the states of the world but has its current states causally disconnected from its proceeding states (or the connection is so weak that it is in fact negligible compared to the influence of the external world) would not be an organism, i.e., a homeostatic dynamic system. This property of organisms implies that two or more states of the world need to be taken into account when we deal with the neurophysiology of the experience of the present moment. The stabilization of this experience requires a certain dynamic structure that links the continuum of the states of the world with the continuum of the states of organism. The corresponding continuous phases can be longer or shorter, but there is a neurophysiological limit to their length, which is aptly called a "time window" or a "window of consciousness." Roughly one to three seconds is needed for a human subject to identify a stimulus (see Pöppel 1989). Although the subjective sensing of the time flow is relatively continuous, the above-mentioned correlating mechanisms function within the frames of the "time window" that is being reconstituted every time anew. The question is: "What happens within the window that enables us to experience continuous flow of the present?" In order to answer this question, we have to look closely at current theories of mental representation. The information that contributes to the experience of the present moment does not come in loose flocks but in the form of representations—the basis of the cognitive content of experience. I shall present just two ideas from the extremely rich field, because they seem to shed light on the issue discussed.

A) Synchronization Based Theory. Thomas Metzinger (1995) departs from an obvious fact that the background of the experience of the present moment is constituted by a large number of physiological phenomena that take place in various locations of the nervous system and are not linked to one another functionally. So, how can they produce anything together?

Metzinger suggests that the mechanism is based on synchronization of neural oscillation in the brain. Referring to the research of Koch and Davies (1994) suggesting that conscious states are linked to the occurrence of neuronal oscillation of 40 Hz, Metzinger advances a hypothesis to the effect of the present conscious states consisting in the synchronization of oscillations of all relevant processes at the level of 40Hz. Actually, his hypothesis is even more general and concerns all representational wholes no matter whether they are conscious or not. "Certain natural representation systems are able to bind internal, spatially distributed individual events which function as feature detectors for them into a representational whole, by coding the perceptive relations between them through processes of synchronization" (Metzinger 1995, sec. 3). Representational coherence and time awareness are explained here at one go. Whatever is responsible for representational wholes also contributes to time awareness and vice versa providing that synchronization at a consciousness related frequency takes place.

Despite its many explanatory advantages, Metzinger's hypothesis is not satisfying. It fails to account for the dependence between the state of the world in the given moment and the conscious experience that occurs in this very moment. It presupposes almost magical connection between the present state of the world and the properties of a higher order characterizing some nervous systems.

B) Simulation Based Theory. In Robert Cummins" (1989) theory of s-representation (simulation based representation) the temporal dimension of experience seems to be given more plausible treatment. Cummins argues that the state of the organism—or, by simplification, the brain state—that represents a state of environment is based on the simulation of this state of environment. This means that the emergence of representation is a process that is always extended in time. For even the simplest state of the environment there exist minimal conditions that must be fulfilled for its simulation to occur. Fulfillment of these conditions needs time; it implies some temporal extension of the corresponding internal processes that may or may not correspond with the physical time of the simulated state. The only temporal limitation for the bundle of simulating processes involved in an experience of the present moment is that they must fit the window of consciousness. This condition, however, allows for considerable time shifts within the window.

One has to be aware that Metzinger and Cummins pose different questions. Metzinger's is: "What gives unity to the dispersed neuronal processes that happen in the same moment?" The answer is: synchronization. Cummins asks: "What mechanism makes neuronal processes represent states of things?" The answer is: simulation. The mechanisms—synchronization and simulation—do not conflict, because they belong to different functional levels. But even when employed together the two theories cannot provide a satisfactory account of the experience of the present moment. Metzinger's theory gives insight into the functional aspect of this experience, while it leaves unexplained its connection with the concurrent state of the world. Cummins's theory puts emphasis on the connection of this experience with the state of the world, but does it in a very complex way that only produces additional difficulties. Cummins reasons in the following way: Representation consists in two processes, simulation and interpretation. Simulations of the states of the world occur in our minds all the time; interpretations emerge only occasionally in regard to whether they are needed by some internal or external activities. Representations are not always useful, and as such they do not result in interpretations (the organism is better off by relying on its reflexes). But when interpretation emerges, it must be interpretation of something. Cummins formulates thus a curious transcendental argument: If simulation is successful and results in a non-trivial interpretation, then the state of things represented by this interpretation exists. (Cummins 1989, p. 105).

Unfortunately, this does not work as an account for the experience of the present moment. It does not determine the temporal location of the state of the world to which the interpretation of the successful simulation refers. If the environment is benign, i.e., it does not change too quickly or too unpredictably, and does not hold any excessive threats, the organism can tolerate quite a great discrepancy between simulation and its interpretation. There is nothing here that could point to the present moment within the continuum of simulation. The flow of the experience is divided into the continuity of simulation with indeterminate objective reference, and discontinuity of interpretation. These objections are exactly symmetrical to one just voiced against Metzinger's theory. Whereas Metzinger gives us the actuality of the experience without reference to the actual state of the world, Cummins secures the reference but without possibility of temporal location of the experience itself.

III. Psychic Events

In order to explain the nature of the experience of the present moment we need to depict some *event*—an entity that is located in time, on the one hand, and carries some content on the other hand. Let us call it a psychic event. My tentative definition may be: *A psychic event is an emergence in a subject a psychological content like thought, perception, or feeling.* One has to be careful and not confuse psychic events with such phenomena as seeing an object, having a thought or the state of feeling cold or warm. A psychic event consists in the emergence of these contents and not in entertaining them.

A good example here is the act of focusing attention on an object. This act shows twofold duality noticed first by Maurice Merleau-Ponty and described in his *Phenomenology of Perception*. The first duality consists in the fact that paying attention to something presupposes that this something has already been given in some other, inattentive way. The second duality consists in the fact that every attentive act is prospective as a whole, because the object is a limit of motion of focusing the attention; it is also retrospective because it will present itself as following the presentation that appears as having motivated the whole process from the very beginning (Merleau-Ponty 1945, 277).

My hypothesis is that these two oppositions: inattention—concentration and retrospection–prospection represent some general feature of psychic events. I suggest that these are based on a repetition of an underlying physical event, whatever it is. Metzinger's time coding by means of neural synchronization is a mechanism whose purpose is to reveal in brain states a common dynamic pattern that becomes—as a property of a higher order—a tool providing information about the environment and the experiencing subject itself. Now, the same mechanisms that stabilize this pattern also make its repetition possible, although the following occurrences of the pattern may be realized by quite different processes in the brain.

Considering their complexity, which probably generates numerous chaotic effects, brain states are in the strict, physical sense, singular. So, if a pattern of oscillation is to result in the constitution of a repeatable higher order property, then the instantiation of the pattern has to fulfill some non-trivial conditions. Among them, as I have mentioned already, are some features of organisms in general, but others, not yet specified, are sufficient conditions that could probably occur only in humans. In order for a psychic event to constitute itself one needs repetition that would be recognized as such. Neural synchronization might well be the physiological mechanism that is responsible for this phenomenon. It is quite possible that the human brain, by tuning together various bundles of neural events at certain frequencies, creates both strict and flexible conditions of the repeatability of very complex physiological processes for which there exists neither a structural nor a causal principle of identity. This is all very speculative though, and I do not think we can do much more to prove or deny the suggestion while staying within naturalistic framework. In the rest of this chapter I shall turn to phenomenology.

IV. Phenomenological Identification of the Present Moment and the Temporal Flow

I have argued that the identification of the present moment needs multiple sources of information. This condition makes its impact on the phenomenal structure of experience. In St. Augustine's famous analysis of recitation in *Confessions* we find a suggestion of a bond existing between the feel of what is now and the intelligible whole towards which the present experience directs itself:

> I am about to repeat a Psalm that I know. Before I begin, my expectation is extended over the whole; but when I have begun, how much soever of it I shall separate off into the past, is extended along my memory; thus the life of this action of mine is divided between my memory as to what I have repeated, and expectation as to what I am about to repeat; but "consideration" is present with me, that through it what was future, may be conveyed over, so as to become past. Which the more it is done again and again, so much the more the expectation being shortened, is the memory enlarged till the

> whole expectation be at length exhausted, when that whole action being ended, shall have passed into memory. And this which takes place in the whole Psalm, the same takes place in each several portion of it, and each several syllable; the same holds in that longer action, whereof this Psalm may be part; the same holds in the whole life of man ..." (St. Augustine, *Confessions* 11,28).

St. Augustine was fully aware that the present state of the world as we see it does not provide sufficient conditions for the experience of the present moment. Our experiential "now" sends us back to a certain "earlier" and forth to a certain "later." The expressions "a certain earlier" and "a certain later" designate the extension of physical time that is necessary to collect all the information determining the current conscious experience, e.g., a perception. The information being gathered within the time gap is much influenced by a person's knowledge.

Yet, this relativization of the problem of experiencing the present moment to knowledge-driven competence to receive the presently available information is still unsatisfying. We need a theory which would determine not only the identification of the moment in the flow of time (via consciousness of the current state of things), but also the principle of the flow itself, that is, the very dynamics of the time experience.

In "William James's Theory of the 'Transitive Parts' of the Stream of Consciousness" Aron Gurwitsch took up the problem of "the intrinsic temporality of an enduring act." He praises William James's theory of transitive parts for overcoming Humean time externalism and placing the temporality of conscious experience within the structure of the conscious act. He also approvingly quotes James's speculation about possible neurophysiological basis of the transitive effects. Gurwitsch admits James's merit in "taking non-perceptual presentation of perceivable objects, meaningfulness of words, experience of relations, and so on, as specific experiences of time, i.e., experiences in which the subject becomes aware of his passing from phase to phase" (Gurwitsch 1966, p.328). For him the main problem with James's theory is that it preserves the main assumption of empiricism, which is the one-dimensionality of time experience. James places the condition of the temporal flow within the act as its transitive part, but although he admits that the transitive parts manifest themselves in different phenomena, they remain indiscernible at the noetic level (speaking in Husserl's terms)—in all cases they are just those properties of acts that enable them to hook on to one another in a flow. The flow itself is presupposed rather than accounted for.

Gurwitsch pays much more attention to variety of manifestations of transitive parts. These are (Gurwitsch 1966, p. 314-316): (a) imageless thoughts, (b) feelings of relations, (c) feelings of tendency and direction, and (d) "psychic overtones," "fringes," awareness in a "penumbral nascent way" of objects and relations. On the one hand, transitive parts are "experiences of connection, continuity and temporality." On the other hand, they are the basis of contents like (a) to (c). Gurwitsch argues that the best framework for explaining the twofold role of transitive parts is his own theory of the field of consciousness. On his theory there are many conjunctions that give consciousness the character of a flow. The conjunctions could not be discerned within the empiricist framework, in which the only possible conjunction is a succession of mental states—succession being best describable in physical terms. In Gurwitsch's field theory, there is more than one way of making mental states flow. "The problem as to whether there is but one or several types of conjunctions leads to a field theory of consciousness. By the latter is meant an investigation into the form or forms of organization prevailing among those facts which at any given moment appear to consciousness" (Gurwitsch 1966, p. 331).

How to explain this variety of conjunctions? Why does the field produce an irreducible plurality of relations between contents? In what respects are the conjunctions different from each other? Why do they shape the time flow differently? The field structure of experience provides an answer. Although Gurwitsch himself does not go very far beyond noticing the field character of necessary relations, it seems consistent with his theory to suppose that the nature of conjunction depends on the position of subsequent contents within the field of consciousness. The term "subsequent" refers in this context to physical time. The field of consciousness organizes itself in physical time, but it is only a necessary and not the sufficient condition of experienced time flow. In other words, there is no direct mapping from physical time flow to experienced time flow. What is needed for the experience of the present moment is consistency of the whole field. This however cannot be accomplished in one run. According to Gurwitsch there are three regions within the field: the theme, the thematic field, and the margin. What seems to happen in the succession of fields are different connections between the contents belonging to all three regions. Although Gurwitsch is not quite explicit here, this is what I suppose he means when he speaks about many conjunctions underlying our time awareness.

The unity of the field within the window of consciousness is a dynamic process involving all the conjunctions. It needs therefore at least double representation of each content within the field. All the field regions belong to each other. It cannot be guaranteed by some external mechanical rule, but the relationship must itself belong to the field. In other words, the unity within the present moment is a property of the field and not the property of the underlying physiological processes. So the contents belonging to different regions have to be in a sense represented within each other. The thematic field must reflect the theme and vice versa. This is especially hard to conceive in case of relation of the two to the margin. But this is exactly where the mechanism of the flow comes to play. The thematic content gets "marginalized" and the marginal content gets linked to the theme and the thematic field.

The dynamic exchange within the field can be further elucidated by referring to Husserl's analysis of temporal consciousness.

V. The Most Radical of Husserl's Reductions

Husserl attempted to merge the conditions of content presentation with conditions of temporal flow in his well-known theory of retention and protention in the *Lectures on the Internal Consciousness of Time*. On this theory the experience of the present moment is grounded in the continuum of present acts which perceive the same content as, on the one hand, more and more sinking in the past (retention) and, on the other hand, as less and less related to its distant origin in the future (protention). The experience of the flow of time is a series of acts where each of them is linked to the former via the same object towards which the subject directs itself. Thus, I see an object as if in its consecutive manifestations where every new manifestation reveals some new perceptional aspects and at the same time repeats the content already perceived. The successive acts follow from one another in a way that resembles the peeling of an onion. But if we want to talk about the experience of the present moment, this onion must have a center: Hence, how is this center constituted? If (as it follows from my previous considerations) this center cannot be defined as a moment in physical time, we can only refer to other segments, properties, and moments of conscious experience.

Husserl's answer goes in two directions. First, he talks about pre-impression as a source of the system of retentions and protentions. Second, he introduces the notion of the pure I which constitutes an ideal correlate of the bundle of acts that are concentrated round the pre-impression.

The theory of retention and protention does not explain fully the phenomenon of the present moment for at least one reason: All these acts which perceive the flow of contents in their coming and going, must have a source whose abode is independent from time. For Husserl, this source is the pure I, which is a highly ambiguous concept. I, the executor of acts and the center of personhood, resists time, but only because it already feels the destructive impact of the temporal flow. Transience and duration are two sides of the same coin, two equally powerful experiences; so there is no theoretical reasons for privileging only one of them as a foundation of the phenomenology of temporality. And there is no reason for maintaining that the pure I should stay untouched by this subjective dynamics.

Husserl convinced himself that the explanation of the "Now" given in conscious experience requires a type of reduction that is the most radical among all the reductions postulated by phenomenology. The purpose of this reduction is to reveal a pure movement of impressions that is ex-centric in regard to the I and to the object-directed experience. The impressional as such is the most primordial transcendence, although a transcendence to be found within the immanence of experiencing.

Following Husserl's remarks, it is difficult to fathom how such a reduction could be effectively implemented. I suppose—linking up to Gurwitsch's framework—that this purpose could be realized via perception of the *marginal* structure of conscious experiences. It would be impossible to understand the essence of temporal experiencing via analysis, even the most subtle, of the content that appears thematically in the experience. Only the margin of experience reveals the object of Husserl's quest, i.e., "the standing-flowing present" (*die stehend-stromende Gegenwart*)—the present moment which, being something solid in time, simultaneously moves in time. The experience of the present moment is not thrown into time, rather, it is time which flows through it.

The reduction proposed by Husserl as an attempt to grasp the nature of the experienced temporal flow consists in moving what is given centrally in its retentional-protentional flow towards the margin; the content is being reduced to pure perception which is no longer the "perception of . . ." It is aimed at revealing the primary dynamics thanks to which any content can appear. This is the deepest *immanence* that is accessible to phenomenological analysis. The pure I appears here not as a problematic entanglement in the reflexive movement of experienced contents (like in the non-egological theory of consciousness endorsed by Gurwitsch), or as a pole of the acts of consciousness, but as one side of the immanent relation between the one *who* experiences and the fact *that* he experiences. This "who" and "that" remain strictly interwoven. "I systematically reduce the concrete flowing present through an 'unbuilding.' I reduce to the primal impressional immanent present of things, to the "foreign to me," namely the immanent *hylē* (sphere of marginalized experience)" (Husserl, *Zeitigung*, MS C 6, 3). This *hylē* that Husserl talks about is something constitutively more primary than the impressional matter that occurs in *Logical Investigations*. This is, in fact, an *Urhylē*, a pre-matter: "Primal *hylē* in its temporization is the ego-foreign core, so to speak, in the concrete present. We would then have to say: in the flow of concrete primal presence, purely immanent time temporalizes itself constantly as primal time, in which [primal] [non-]individual being exists; thereupon, we postulate as valid the pure flow of experience, the first "transcendence" over and against the primal-impressional flowing present" (Husserl, *Zeitigung* MS C 6, 5).

The hyletic field passes continuously through immanent time and is thus the hyletic core material of the primal-impressional sphere. In other words, Husserl tries to understand the nature of perception as perception. Beneath any given content, it emerges from the dynamics of the present moment; and this dynamics is a movement of the primordial impressional matter itself: "In this 'unbuilding' we return to the primal-impressional present, which leaves valid exclusively the purely perceived and reserves in bracketed form that which is not purely perceived, withholding in this from pure perception 'perceiving' and all other consciousness, and that as non purely perceived in any case, and as perhaps not perceived at all" (Husserl, *Zeitigung* MS C 6, 4). The unclear formulation in the second part of this sentence I would interpret in the following way: The bottom level of perception is not a form of its corresponding objectivity, but the pure movement of impressions; in this mode of consciousness, strictly speaking, nothing is yet perceived. But what does Husserl mean when he talks about unbuilding (*Abbau*)? Certainly not a conceptual analysis separating the experiential components or any real disintegration of the experience. It is the phenomenon of pure receptivity, which is elucidated by this part of Husserl's study.

VI. The Problem of Combining the Phenomenological and the Cognitive Approaches

Husserl's analysis does not say anything about the *realization* of the revealed content structure. On the other hand, its great advantage lies in the way it tackles the constitution of time as dependent on both sides of experience, subjective and objective alike. It transcends this opposition by showing how the movement away from the thematic content towards the marginal content converges with the position of the I in the field of sensuous receptivity. This I is very far from the Cartesian cogito. It emerges from the dynamics of the temporal experience that is effected by a complex mechanism of horizons. The I who operates with consciously experienced contents is a content for itself; whereas the I who experiences its *now* reveals its specific being as a correlate of the possibility of receiving impressions.

It is tempting to conclude at this point by saying that this ideal receptivity or sensuousness is nothing else but a biological property of the human body, and the tensions postulated by Husserl within the field of experience result from incommensurability in the functioning of separate units of human brain. One could say that from this heterogeneous bedrock a relatively coherent self-representation emerges that somehow contains all these incommensurabilities. In cognitivism, this representation relies on a brain model of one's own body. This model might be said to determine the activity of the living body as a receiver of impressions. The sense of time and, above all, the sense of the present moment would then be associated with the transaction between the abstract, timeless model of one's own body and the physical temporal continuum that is particular to the processes occurring in this body.

This approach, apparently promising, is, in fact too close to the old-fashioned functionalism and its solipsistic consequences (in the style of early

Putnam and Fodor). It ignores the fact of receptivity as a possible foundation of the sense of time. Here, any process happening inside the body could give rise to the consciousness of time, and the experience of the *now* in particular. This view violates the reasonable condition I have formulated before, according to which the experience of the present moment should be conceived as intentionally directed towards the state of the world in this very moment.

Husserl goes in the more promising direction, although his model does not contain any empirical data. On his analysis, the body is not presented as an abstract model built into a sensing organism, but as an individual concept of a living body that is given with and in the stream of experiences (C 6, p. 7). This individual concept (that is, the concept of one's own body) is revealed in transcendental analysis, but only as a transcendental possibility. It is impossible to perceive it effectively, for it is always implied by any constituted human experience. The sphere of functioning of this living body can be seen only through the layers of already constituted knowledge. The consciousness of the present moment understood as the consciousness of impressional receptivity that is particular to the living body, existing here and now, can be created only as a "result" of the immanent unbuilding of experience to the level of *hylē*. It cannot be done effectively, for we always experience something already constituted. This is, I think, what distinguishes Husserl's transcendentalism from any cognitive analysis where all the constituents of experience must effectively be shown. Husserl is aware that the reduction is performed on the living, already accomplished experience. Thus, in the strict sense, we do not experience hyletic data, nor do we fulfill retention and protention as separate acts, nor do we "marginalize experience" in the way as, for example, we can marginalize at will the range of our perception by half-closing our eyes or by concentrating on the edges of our perceptional field.

With all the due reservation coming from methodological grounds, one can search for compatibility between the transcendental method of explaining the temporal experience adopted by Husserl and the theories of neurophysiological mechanisms of consciousness. It is true that the Husserlian analysis deals with possible experience, whereas the mechanisms investigated by cognitive psychologists and neurophysiologists point to the conditions of its realization. It is also true that individual sentences in phenomenology and in the psychological-physiological theories are mutually untranslatable. Compatibility, however, is something else than successful translation. It manifests itself not on the level of separate statements but on the level of postulated properties of a higher order, as well as other ontological and epistemological assumptions. As I already mentioned with respect to Metzinger's theory, the synchronization of the brain's functioning at the level of about 40 Hz is probably supported by the mechanism that represses other frequencies. On the phenomenal level, it is represented as the selectivity of consciousness. But this repressing does not amount to nullification. Apart from synchronization, there are other causal connections that tie brain processes independently of the links based on synchronization. The effective absence of certain contents does not yet indicate that they leave no traces in the form of our experiences. These are precisely the traces phenomenological analysis tries to discover.

Conclusion

I have argued that the experience of the present moment manifests itself in the emergence of contents (psychic events). They can be identified by both scientific and phenomenal methods. The higher order properties that are assumed by cognitivist theories can be thought of as grounded in transcendental conditions of experiences understood in the Husserlian sense. Speaking from the cognitivist perspective, I postulated the repetition of physical events as the basis of psychic events as well as to the mechanisms of selection, simulation and synchronization from which conscious experiences emerge. I confronted these conclusions with Husserl's theory of retention and protention and the idea of the immanent deconstruction of the experience leading to the pure movement of impressions and their correlative subjective structure, the pure I. Also I discussed Gurwitsch's theory of the field of consciousness as a good framework for describing the experience of the present moment. This description is based on the notion of content marginalization, which in turn presupposes both synchronic duplicity and diachronic repeatability of content representation. This is where the link comes to cognitive theories which for different reasons should also deal with the problem of duplicity and repeatability of mental representations.

The cognitivist and phenomenological approaches to the experience of the present moment cannot be straightforwardly put together but they are not incompatible. The transcendental (constitutive) analysis describes conditions of possible experience, while cognitivist theory describes conditions of realization of these possibilities. Yet the full analysis of the experience of the present moment must resort to such transcendental conditions for which there is no empirical theory of realization. Such levels (immanent structures) of experiencing as receptivity, the movement of sensual impressions, the pure I, and the marginalization of experience will probably never be explained in terms of conditions of realization. This cannot be said with certainty, however. The cognitive theories change rapidly and it is difficult to say now what philosophical use can be made of the future findings.

References

Augustine. 2001. *Confessions*, translated by E.B. Pusey, Everyman's Library, A. A. Knopf, Inc.

Cummins, R. (1989. *Meaning and Mental Representation*, Cambridge, Mass.

Gurwitsch, A. 1966. William James's Theory of Transitive Parts of the Stream of Consciousness, in A. Gurwitsch, *Studies in Phenomenology and Psychology*, Evanston.

Husserl, E. 1966. *Vorlesungen zur Phänomenologie des inneren Zeitbewußtseins*, in *Gesammelte Werke*, Bd. X: *Zur Phänomenologie des inneren Zeitbewußtseins* (1893-1917), ed. R. Boehm, Den Haag.

—. (undated). *Die C Manuscripte* (Manuscript C6), Husserl Archiv in Leuven, in preparation by D. Lohmar.

Koch, C., and J. L. Davis. 1994. *Large-Scale Neuronal Theories of the Brain*, Cambridge MA.

Merleau-Ponty, M. 1945. *Phénoménologie de la perception*, Paris.

Metzinger, T. 1995. Faster than Thought. Holism, Homogeneity and Temporal Coding, w: *Conscious Experience*, Thorverton, UK. Also available on the Internet: www.zynet.co.uk/imprint/online/Metz1.html

Pöppel, E. 1985. *Grenzen des Bewusstseins überWirklichkeit und Welterfahrung*, Stuttgart.

Chapter 6

Field Theories of Mind and Brain

Jeff Yoshimi
University of California, San Diego

Abstract

Aron Gurwitsch's Gestalt-inspired "field theory of consciousness" was introduced in the same period as Wolfgang Köhler's theory of "electrical brain fields." I consider parallels between these theories, drawing on results that have emerged in the last five years. First, I consider the claim that fields of consciousness supervene on electro-magnetic fields in the brain, then I outline Gurwitsch's field theory of consciousness, and finally I consider how the structures described by Gurwitsch might relate to structures in the brain's electro-magnetic field. Along the way, I expose a dogma (that qualia are paradigmatic conscious states) and develop an interpretation of Gurwitsch's field theory.

The electro-physiological methods and phenomenological analysis of psychical behavior... can reciprocally confirm one another (Aron Gurwitsch, 1940).[1]

Introduction

Aron Gurwitsch was the first phenomenologist after Husserl to think deeply about issues in empirical psychology.[2] He began under Carl Stumpf around 1919, worked with Gelb and Goldstein on brain-injured soldiers in the 1920's, and wrote a dissertation on Gestalt psychology and phenomenology (1929). He not only worked within psychology and phenomenology, he actively theorized *between* them: among other things, he saw in Gestalt psychology an "incipient phenomenological reduction," he correlated brain pathologies with failures of

[1] From *Studies in Phenomenology and Psychology* (hereafter SPP), 83.

[2] On Husserl's view of the relation between phenomenology and psychology see his *Phenomenological Psychology* as well as Gurwitsch's discussion of these lectures in *Phenomenology and the Theory of Science* (hereafter PTS). On Gurwitsch's view of the relation between psychology and phenomenology see Embree (2003).

L. Embree (ed.), *Gurwitsch's Relevancy for Cognitive Science*, 111-129.

perceptual abstraction and ideation, and he applied the Gestalt concept of "good continuation" to systems of co-referential noemata.[3]

But Gurwitsch's most original contribution—his "field theory of consciousness"—makes little mention of neuroscience, despite its obvious relevance.[4] Indeed, Wolfgang Köhler, the leading Gestaltist (also exiled in America at that time), developed his own theory of "electrical brain fields" with explicit reference to mental-physical "isomorphism."[5] In what follows I consider these long-standing but largely unexplored parallels, drawing on results that have emerged in cognitive and neuroscience since the 1960's, and in particular in the last five years.

I. Field Supervenience

In this section I pursue a broadly Gurwitschean strategy for linking conscious states with brain states. In *(A)* I argue that it is a mistake to begin one's study of mind and brain by focusing on "parts" of conscious states (e.g., "qualia"), for, as Gurwitsch and others show, the total field of consciousness is always primary. In light of this I articulate *global* mind-brain supervenience in *(B)*, which links changes in total conscious state with changes in total brain state. In *(C)* I review evidence for a more specific hypothesis, "field supervenience," according to which changes in what Gurwitsch calls the field of consciousness entail changes in the brain's electro-magnetic field.

(A) The Mereological Dogma. A widespread assumption in philosophy of mind is that "qualia"[6] (in particular "color qualia") are paradigmatic mental states. More generally *parts* of mental states are taken to exemplify mental states in their totality.[7] But this is a dangerous assumption, for it conceals important preparatory studies that should not be shrugged off lightly. After all, "color qualia" are integrated into broader mental states which encompass colored *and*

---

[3] See *The Field of Consciousness* (hereafter FC) p. 168 and SPP, p. 103; SPP, 359; and FC, p. 217, respectively. Gurwitsch's other contributions at this interface include detailed studies of James and Piaget vis-à-vis phenomenology. See, e.g. SPP, chs 8 and 12, and FC part 1, ch. 1-2.

[4] There is some discussion in FC, 55, n. 74 .

[5] According to which "the structural characteristics of brain processes and of related phenomenal events are likely to be the same" (Köhler, 1959, p. 22).

[6] "A quale is an introspectible and seemingly monadic property of a sense-datum. For example, the qualia of a visual sense-datum of a rose would include the experienced redness, and the qualia of an olfactory sense-datum of a rose would include the sweet-ness of the scent" (Eliasmith).

[7] The ubiquity of this assumption is manifest in the difficulty of finding authors who *do not* make it. Rarely are mental states referred to in their undivided totality. "Percepts," "sensations," and "qualia" are the standard exemplars of mentation.

textured objects, awareness of the body, non-perceptual moods and tinges, and so on through a rich and complicated structure that phenomenology has articulated in great detail. I shall call the notion that one can *begin* one's study of mind by focusing on *parts* of conscious states the "mereological dogma."

In the analytic tradition Rudolph Carnap recognized this dogma and its dangers, scolding his fellow philosophers for beginning with parts, since in doing so "[they do] not take the given as it is, but [rather take] abstractions from it....as basic elements" (§67). Instead, Carnap advocated beginning with what is "epistemically primary," namely *total* mental states—what he called "elementary experiences" or more famously, "the given": "we have to proceed from that which is epistemically primary, that is to say, from the 'given,' i.e., from experiences themselves in their totality and undivided unity" (§67).

Among phenomenologists it was Gurwitsch who described "experiences in their totality and undivided unity" in the most detail and who most carefully avoided what I am calling the mereological dogma. In fact he arguably denied that parts *could* be identified within total fields of consciousness.[8] From his early work in the 20's through to his last essays, he always had total fields in view and was careful to articulate other phenomena in relation to these totalities.

Avoidance of the "mereological dogma" is especially pressing for the project of correlating conscious states with brain states. After all, if one begins with *parts* of mental states, one has to outline what might be called a "structural neuro-phenomenology," which takes account of how conscious states divide into parts, how brain-states divide into parts, and how these two kinds of part relate to one another. This is an important project, and one for which Gurwitsch provides valuable insights (cf. Part II following), but it is not the place to begin in a systematic development of neuro-phenomenology.

As an example, consider John Searle's (1983) parsing of intentional states into psychological force and representational content (in early Husserl: quality and matter). In desiring a cup of water, for example, "desire" is the psychological force and "that I obtain a cup of water" is the representational content. How might this division be accounted for in neurological terms? There ought to be *some* story to tell, but this is a difficult question to begin with. After all, the question of how the distinction works in the mental domain is itself tricky, involving an obscure parsing of intentional states into force and content, and concealing the further question of how intentional states relate to wider mental states. Then, presuming we answer these questions adequately, we must say something about how these various part-structures map on to correlated part-structures in the brain.

---

[8] Insofar as Gurwitsch allows for "parts" of a field of consciousness they are "whole-parts," which are "essentially determined and qualified by the functional significance they have with respect to each other and hence, for the whole of the Gestalt contexture into which they are integrated" (FC, 149).

On the other hand, what Gurwitsch calls the "field of consciousness" has an intuitively plausible neural correlate: a total field of activity in the brain. This may or may not be right (evidence is reviewed shortly), but it is certainly a more straightforward claim, and a useful basis on which to approach more complex structural questions such as the one about how force and content relate to neural structures. The reason is that fields of consciousness are *wholes*, and so too with fields of activity in the brain. The question of how fields of consciousness (and brain states) divide into parts remains important, but is initially bracketed in order to focus on the simpler question of how global mental states relate to global brain states.

Two objections are worth considering. First, one might object that in science one routinely begins with simple constituents rather than complex wholes. One would not tell the physicist: you ought to begin with complex forces, not simple ones. Or the chemist: you ought to begin with tables and chairs, not molecules and atoms. Response: In each of these cases the low-level analysis is only made possible against a tacit but well-worked out mereology. In physics an analysis of simple forces relies on an antecedent understanding of what force-vectors are and how they can be decomposed into components. Molecules are useful because the way tables and chairs comprise them is well-understood. But the parallel mereological theories in neuroscience and phenomenology have yet to be fully worked out. At the very least we ought to have the question of part-whole relations in mind as we seek low-level neuro-phenomenological correlations.

Second, I have become convinced that binocular rivalry (see Leopold and Logothetis, 1999), where incompatible stimuli are separately presented to a subject's two eyes (resulting in a kind of "flicker" between two percepts), is a useful paradigm in the search for neural correlates of consciousness. The reason is simple: it is an experimentally tractable set-up which allows researchers to seek direct physical correlates of a changing conscious state. But my general point stands: researchers ought to bear in mind that the "flickering" percepts of binocular rivalry are parts of wider mental states, and that ultimately we will have to have a story to tell about how the relation between percepts and global conscious states maps on to a parallel relation in the brain.

(B) Global Supervenience.[9] A plausible and relatively minimal way of linking global conscious states with global brain states is by supervenience. Supervenience is an asymmetric determination relation, where "supervenient properties" (in a set *A*) are determined by "subvenient" or "base properties" (in a "sub-

[9] In the supervenience literature "global supervenience" refers to a relation between properties of possible worlds (hence "global"). "Global" is here used in a different sense, to refer to conscious states and brain states in their undivided totality, as opposed to abstracted parts of either.

venient base" set *B*) but not vice-versa.[10] In other words, as the supervenient *A* properties vary so too do the base *B* properties, though changes in base *B* properties do not entail changes in supervenient *A* properties. For example, if the ethical supervenes on the physical then a change in ethical situation entails a change in physical situation. If the world goes from being good to bad then something *must* have changed physically. But a change in the physics of the world (e.g. the removal of a few atoms) does *not* entail a change in its ethical character.

My claim in this section is that:

> (1.1) Global conscious states supervene on global brain states.

Change a person's global conscious state and his or her brain-state will change as well. But not necessarily vice-versa: there are indefinitely many ways a brain-state could be incrementally changed without that change having any affect on consciousness.[11] Each conscious state has (at least in principle) multiple cerebral realizations.

Note that *A* and *B* in (1.1) comprise possible states of persons at times—possible conscious states and possible brain states.[12] Supervenient states in *A* are global Carnapian mental states. No parts, no divisions (yet), just the totality of consciousness for a person at a time. Subvenient states in *B* are global brain-states. What particular aspect of the brain is relevant to consciousness is left open. We do not yet know exactly what in the brain correlates with consciousness, but we assume that *something* does.

---

[10] More specifically, *A* properties supervene on *B* properties iff a difference in *A* properties entails a difference in *B* properties. See Kim (1993), chs. 4-7 and Savellos and Yalçin (1995) for comprehensive discussion. The concept of supervenience is highly ramified. This definition captures what McLaughlin calls the "core idea" of supervenience (1995, p. 18), namely, that base-properties *determine* supervenient properties. It does not, however, capture ontological dependence, which is usually taken to be essential. A definition which *does* entail both determination and dependence (at the cost of unwieldiness) is this: "A states supervene on B states iff necessarily, for any A-state there exists a B-state such that, necessarily, any individual in that B-state will also be in that A-state." For more on the logic of supervenience see my "Mapping Supervenience."

[11] So supervenience is not the kind of one-to-one relation that Wiggins (this volume) describes in relation to the constancy hypothesis (which Gurwitsch was an outspoken critic of).

[12] *A* and *B* are sets of states, that is, mutually exclusive properties. In this context we have two sets of mutually exclusive properties—possible brain states and possible conscious states—and our domain comprises people at times. A single state in each set will apply to each person at each time—people are always in some conscious state in *A* and some brain state in *B* (where we imagine, say, a "null state" in *A* for dreamless sleep and coma).

One advantage of supervenience is its metaphysical innocence. It is, for example, neutral on the mind-body problem. As Jaegwon Kim (1988) put it: "...mind-body supervenience states the mind-body problem—it is not a solution to it." Supervenience is compatible with mind-brain identity theories, for example, but also with dualist and even idealist perspectives.[13] Given the seeming intractability of the mind-body problem I find this to be an attractive feature of supervenience—it allows us to make substantial and (at least in principle) testable claims about the covariance of conscious-states and brain-states without answering the question of their real relation.

Global supervenience (1.1) is so well supported it has hardly been the subject of scientific inquiry. It is given inductive support whenever a change in supervenient conscious-state is observed to co-occur with a change in subvenient brain-state. But of course, one's conscious-state and brain-state are both almost *always* changing. Constancy in either domain, *lack* of change, is nearly impossible to observe. Nevertheless, (1.1) could be falsified in at least two ways, insofar as supervenience implies (a) dependence of consciousness on the physical and (b) determination of consciousness by the physical (see footnote 10). (a) could be falsified by a disembodied consciousness or spirit, a conscious state *not* co-occurring with a brain-state. (b) could be falsified by a single brain state supporting multiple conscious states. Neither situation has ever been observed in a non-problematic way.

(C) Field Supervenience. I now sharpen (1.1) by replacing "global conscious states" with "fields of consciousness" and "global brain states" with "electro-magnetic fields in the brain" yielding:

> (1.2) Fields of consciousness supervene on electro-magnetic fields in the brain.

That is, a change in a person's field of consciousness entails a change in the electro-magnetic (or "EM") field occurring in that person's brain.[14] Again this co-variance is asymmetric: a change in my brain's EM field (in particular a slight change) does not entail a change in my field of consciousness.

In focusing on "fields of consciousness" I incorporate Gurwitsch's specific, physics-inspired field metaphor (a metaphor integrally linked with Gurwitsch's

---

[13] The fact that supervenience is largely neutral with respect to the mind-body problem is shown by constructing dualistic, physicalist, and idealist models which satisfy the relation. See Yoshimi, 2001, ch. 3.

[14] Strictly speaking, the electro-magnetic field is pervasive. So to speak of "a" field of EM activity in the brain is somewhat misleading. What I mean by the phrase is an especially strong and complex region of the pervasive EM field, localized inside the human skull, generated largely by the movement of ions into and out of cells in neural transmission and conduction.

broader theory of consciousness, discussed in part 2).[15] In focusing on electromagnetic fields in the brain I have further specified the subvenient base *B* of (1.1). Recall that, in this context, a subvenient base is a set of possible states of a system. Previously these were global brain states, possible total states of a brain. I now focus specifically on the EM fields generated by (among other things) spiking neurons. Glial cells, blood cells, cerebro-spinal fluid, and other substances are left out of consideration.

The concept of an "electrical brain field" was first articulated by Wolfgang Köhler (1949, 1951, 1952), who emphasized the fact that this field extends beyond the cell membranes of individual neurons: "In referring to nerve impulses, we commonly think of events which occur in fibers [neurons] and in their most immediate environment. From these impulses we distinguish their fields which-...spread farther...current flows partly in the fiber, but necessarily also through the surrounding tissue.... The field pervades the tissue as a continuum" (1951, p. 227). Several recent theorists (Pockett 2000, John 2001, McFadden 2002a,b) have resurrected the concept of an electrical brain-field in its relation to consciousness. These thinkers diverge in important ways, but agree that the brain's EM field correlates with consciousness.[16]

What reason is there to believe that this particular subvenient base is the right one? What reason do we have to move beyond (1.1) in this specific and controversial way? Some personal history is instructive here. My initial reading of Gurwitsch led me to suspect that the kind of structured and evolving field of consciousness he described was realized in a similarly structured and evolving field of activity in the brain. In light of this, I was heartened to discover that people like Köhler had followed up on this hunch. I remain optimistic, though my optimism has been dampened by careful reading of the relevant literature.

The situation is roughly this. Good evidence has been provided for (1.2), that is, for the claim that a change in consciousness entails a change in the brain's electromagnetic field. But most of this evidence speaks equally for the claim that consciousness supervenes on neural activity (because changes in the EM

---

[15] Gurwitsch taught physics at Harvard during the war and published a paper in the *American Journal of Physics* in 1946 ("Algebraic Study of Lenses," volume 14). More importantly, he was clearly aware of the way the field metaphor from physics was being used by Gestalt authors such as Köhler and Lewin (see SPP, p. 62).

[16] John emphasizes a "comparator" function of the field, whereby the present is compared with the past, and describes the field in terms of "clouds of ions" and "heterogeneous charge densities." McFadden emphasizes the information content of the field, as well as its ability to affect motor neurons and thereby communicate with the external world. Also note that McFadden (2002b, p. 48) emphasizes that it is *not* the total EM field that correlates with consciousness; rather "only a tiny component of the information held in the brain's EM field can correspond to consciousness." So McFadden would further restrict the subvenient base of (1.2) to encompass only some *part* of the EM field. Pockett focuses soberly on correlations between consciousness and the EM field.

field imply changes in neural activity).[17] So the evidence in favor of (1.2) for the most part also supports the following "neural supervenience" claim:

> (1.3) Fields of consciousness supervene on patterns of neural activity in the brain.[18]

This leads one to seek evidence specifically in favor of one or the other subvenient base (electro-magnetic fields or patterns of neural activity). And here I think the situation is roughly balanced. Some evidence speaks in favor of (1.2) more so than (1.3), which suggests that the EM field is the subvenient base of consciousness and that neural activity is a causal antecedent which could in principle be eliminated without affecting consciousness. But there is also important evidence from the 1950's (Lashley et al., 1951, Sperry et al. 1947, 1955) in favor of (1.3) and *against* (1.2), which suggests that the EM field is a mere epiphenomenon, and that it is neural activity that is the real subvenient base of consciousness. In part because of Sperry's and Lashley's experiments (which had a devastating effect on Köhler's views), this is the dominant position today. Of course it is also possible (and in fact rather likely) that *both* claims are true—that consciousness supervenes on neural *and* electro-magnetic activity. For rhetorical force and conceptual clarity I separate these claims in this section, leaving intermediate positions as an open possibility.

Before considering this evidence, it will be useful to review the kinds of intuitions that motivate (1.2) as against (1.3). Of course intuitions by themselves provide little in the way of evidence, but they play an important role in theorizing (motivating experimental designs, for example) and, left unstated, can be dangerously misleading (tempting us to confuse intuitions with actual empirical support). So we ought to be explicit about them.

The major intuition motivating something like (1.2), which played prominently in the Gestalt theorizing of the 50's (and continues to be influential today[19]), is that of phenomenal continuity. Both the electro-magnetic field and the conscious field are continuous wholes. All the various elements of consciousness at a time are unified or "bound" into one total awareness at that time. But patterns of neural activity are spatially discrete, occurring at localized points

---

[17] But *not* vice-versa; changes in neural-activity do not imply changes in the EM field (at least not in principle). The means that EM fields supervene on neural activity, so that insofar as consciousness supervenes on EM fields it also supervenes on neural activity (transitivity of supervenience holds here, though it does not always hold; see my "Mapping Supervenience").

[18] How to define "neural activity" is itself controversial. I will assume that "neural activity in the brain" refers to a distribution of instantaneous spiking rates over the entire population of neurons in a brain at a time.

[19] See McFadden, p. 44 ff, John p. 198ff, and Pockett pp. 10-11. For general discussion see Brook, 2001.

inside the brain. Hence the continuous EM field is taken to be a more likely correlate of consciousness than a pattern of neural activity.

There are also intuitions about the "granularity" of consciousness vis-à-vis the brain's EM field. It seems impossible to parse a field of consciousness into more than hundreds of parts. There are the objects in my visual field, my sense of my body, my thoughts, etc. On the other hand, there are billions of active neurons in the brain. An intuitively compelling line-up between the two fields could be obtained by partitioning the EM field into tens or hundreds of spatial regions, one for each datum in the field of consciousness.

Also compelling is the very use of a "field metaphor" to bridge the two domains—phenomenology and neuroscience. There is something intuitively plausible about a "field of consciousness" evolving in parallel with an EM field. In both cases we have a unified but structured totality which changes over time according to laws. In my dissertation I call the field metaphor a "bridge metaphor" which harmonizes our intuitions about mind and brain.

To begin to think about (1.2)'s empirical content, let us consider what it would take to falsify it. Recall that supervenience claims can be falsified in at least two ways, insofar as they entail (a) dependence and (b) determination. (a) Fields of consciousness depend on EM fields in the brain. So one could falsify (1.2) by effectively shutting down the brain's EM field and showing that a field of consciousness persists. For example, one might encase all the neurons of the brain in a shielding material and thereby prevent the field from extending beyond individual neurons. If consciousness persisted in such a case, (1.2) would be falsified. (b) EM fields determine fields of consciousness, in that changes in consciousness imply changes in the EM field. Hence (1.2) would be falsified if one were to induce a constant EM field (using, say, coils and a feedback system) in the brain of a person (accurately) reporting changes in their consciousness.

But these situations are science fiction of the type philosophers of mind (insofar as they are confined to armchairs) engage in. None of the proposed falsifications above is feasible in practice. In light of this, more realistic experiments testing the *correlation* between consciousness and brain states have been carried out.

Unfortunately, most of these experiments provide equal support for (1.2) and (1.3). EEG and MEG, which directly measure the EM field of the brain (and hence neural activity as well), have been used to correlate specific field patterns with specific kinds of consciousness. Waking, deep sleep, and drowsiness, for example, each manifest a characteristic EEG pattern (see Pockett, 2000). Electrophysiology, which directly measures spiking neurons, has shown that neural activity (and hence EM field activity) correlates with various perceptual changes in monkeys (see, e.g., Leopold and Logothetis, 1999). Brain imaging techniques, which indirectly measure neural (and hence, EM field) activity via blood metabolism and oxygenation have also shown correlation between changes in conscious state and changes in neural/EM state (see, e.g., Polonsky et. al, 2000). Anesthesiology studies (Fiset et al. 1999, John et al. 2001) using combinations of the above techniques have identified large scale changes in the EM field (via EEG) and neural activity (indirectly via PET) that correlate with loss of consciousness under general anesthetic and return of consciousness as the anesthetics wear off.

In light of the indeterminate nature of these studies, one is led to seek evidence specifically in favor of one or the other subvenient base. I begin with field supervenience (1.2). There is some evidence that changing the brain's EM field (as opposed to neural activity) affects consciousness. It has been shown, for example, that transcranial magnetic stimulation or "TMS" (Hallett, 2000 and Ruohonnen et. al 2000), whereby a powerful magnetic field is used to non-invasively alter the global EM field in the brain, produces "a range of cognitive disturbances in subjects including: modification of reaction time, induction of phosphenes, suppression of visual perception, speech arrest, disturbances of eye movements and mood changes" (McFadden, p. 30). Still, magnetic stimulation affects neural activity as well as the EM field, which leaves open that it is the change in neural activity which correlates with the change in consciousness. Indeed TMS is often used precisely as a means of inducing neural activity.

McFadden and John have theoretical arguments in favor of (1.2) as against (1.3). McFadden enumerates seven predictions made by his version of the field theory of consciousness (2002a, pp. 31 ff), each of which is better-supported by field than neural supervenience. Among these is the fact that "increased complexity of conscious thinking should...correlate with increased complexity of the brain's EM field." And indeed, (John et. al. 2000) supports this notion, insofar as consciousness as opposed to deep (and presumably dreamless) sleep is associated with higher mean frequency in the EEG across all frequency ranges, indicating a more "complex" EM field. But this is hardly conclusive evidence that consciousness supervenes on the EM field alone. After all, these changes in mean frequency are themselves the results of changes in underlying neural activity. John's theoretical arguments are more complex—involving "negative entropy" in the EM field and the inability of a neural network to evaluate this entropy.

What evidence is there in favor of neural supervenience (1.3) and *against* field supervenience (1.2)—i.e., evidence for the claim that the EM field is a mere epiphenomenon? The relevant studies were carried out in the late 40's and early 50's, by the now-famous neuroscientists Karl Lashley and Roger Sperry. In these experiments the electrical fields generated in the cortex of a cat or monkey were altered, and it was shown that some learned perceptual behavior would be unaffected by the change.[20] In addition to these experiments, Lashley surveys

---

[20] Lashley's approach was to distort the electrical fields of two monkeys by connecting diverse regions of one's cortex with gold foil and by short circuiting local regions of the

a variety of theoretical reasons for doubting the field hypothesis (e.g., the fact that variations in the field caused by fluid movements produce no noticeable affect on consciousness).

There are reasons to be skeptical of these experiments and arguments (cf. Pockett's discussion on p. 119) and, to his credit, Lashley is forthright about many of these reasons (e.g. the fact that the relevant effects of the inserted gold strips and pins were unknown). But the main problem, from a phenomenological perspective, is the fact that such simple behavioral tests were used. It may be that the consciousness of the animals *was* affected but not strongly enough to prevent them from successfully completing their discrimination tests and thereby receiving the sought-after food-morsel. After all, consciousness is notoriously underdetermined by behavior—any given behavioral pattern is consistent with a wide range of conscious states. Again, Lashley and Sperry, to their credit, are forthright about such limitations, and in fact Sperry points to the need (unrealized in the 50 years since) for more "definitive experiments" (1954, p. 50).

So the jury is still out. One hopes with Sperry for more definitive experiments. It is worth noting that whichever way the cards fall Gurwitsch is at least potentially relevant to cognitive science. If (1.3) prevails, then a very nice and intuitive linkage between the two fields is preserved. If (1.2) prevails, then some of the intuitive linkages are lost but substantial relations between the field of consciousness as described by Gurwitsch and a global pattern of neural activity might yet be identified. And of course a mixed contribution of neural and EM activity also remains plausible.

II. Field Structure

Near the beginning of his *opus*, Gurwitsch says: "For a field theory of consciousness there arises the task of investigating the articulation of the total field of consciousness and of bringing out the patterns and forms in which co-present data are organized with respect to each other" (FC, p. 2). Indeed, the bulk of *The Field of Consciousness* is devoted to this task of describing the structure and dynamics of the conscious field. Having framed the claim (1.2) above that conscious fields supervene on EM fields, I will formulate more interesting (and con-

---

other's with gold pins. Each monkey had been trained on four visual discrimination tests (e.g., choosing a red vs. a green square), and it was shown that pre- and post-operative performance was essentially identical. Sperry's experiments were similar—he inserted tantalum (which is biologically inert) in a criss-cross pattern across each cat cortex and showed that the cats performed as well on visual discrimination tests after the operation as before. He did a similar experiment with monkeys and motor coordination tasks, although this experiment was confounded by the fact that control animals with relevant areas *completely* removed also performed well on the motor tasks.

troversial) claims about *structural parallels* between the two fields, taking Gurwitsch's field theory as my lead.

(A) Overview of Gurwitsch's Field Theory.[21] Gurwitschean field theory can be characterized in terms of six basic theoretical commitments.

1. The field of consciousness, says Gurwitsch, comprises "co-present data" which are variously organized and related to one another—the multifarious contents or parts of total awareness at a time.

2. Of these, some are prominent and coherently organized at the focus of attention or in Gurwitsch's terms, the *theme*. These contents are organized according to Gestalt law so as to detach themselves from a wider background and present themselves as coherent wholes. Examples include a group of dots seen as a group, a face, a mathematical theorem, or a memory—assuming in each case that the relevant experiential "Gestalt" is at the focus of attention.

3. From these inauspicious beginnings (the relevant phenomenology was standard even in his day), Gurwitsch develops a comprehensive theory of the structure and dynamics of the conscious field. *Relevancy* is the central organizing principle in this theory.[22] Though Gurwitsch does not offer an explicit definition of this concept (regrettable given its central importance), he does say enough that a definition can be extracted. To begin with, note that "relevance" is a relational term, where the relata are "data" in a conscious field. The basic intuition is that data in a field are relevant to one another if they are experienced as being somehow connected or associated with one another. A less circular definition is as follows: Co-present data *a* and *b* in a field of consciousness are *relevant* to one another iff varying *a* impacts the perspective under which *b* is experienced.

To take a Gurwitschean example, suppose you are contemplating a historical event, say the First Crusade. You can think of the First Crusade as an event in the Middle Ages, as an exemplar of Romantic chivalry, or as an antecedent to the holy wars of the 21st century. In each case the perspective under which you ponder the First Crusade is different; in each case your thoughts of the Crusade are differently "tinged." These data are *relevant* to the attentional theme. Thus, varying our context-awareness from, say, "event in the Middle ages" to "exemplar of Romantic chivalry" impacts the perspective under which "the First Crusade" is experienced.

But thoughts of the First Crusade are hardly sufficient to dominate consciousness. As Gurwitsch says in the context of mathematical thinking, "When we are dealing with a theoretical problem, more than the problem alone is given

---

[21] An early statement of the theory is in Gurwitsch (1929). The mature statement is Gurwitsch (1964). Though Gurwitsch is a clear writer, the details of his field theory are controversial, so what I say should be taken as first steps towards a general interpretation and development.

[22] See Embree, "Three Species of Relevancy in Gurwitsch," this volume.

to consciousness" (FC, p. 1). As you ponder this historical event the world remains in the background, on the "margins" of your conscious field. The desk you are sitting at, a pad of paper, and "through the window, a segment of scenery outside the house," remain present to mind as you pursue your historical ruminations. So too with your body—you have some awareness of your shifting posture, of the clothes you wear, of your grumbling stomach, etc., as the historical musings unfold. These contents are *not* relevant to the theme. Varying the color of the desk or adding a flock of birds in the background does *not* affect the perspective under which "the First Crusade" is experienced.[23]

4. On the basis of his concept of relevancy, Gurwitsch sorts non-attentional data into two classes: data relevant to and data irrelevant to attention. That is, he distinguishes two forms of non-thematic background consciousness. First, a "thematic field" of data relevant to and so covariant with the theme, a kind of Jamesean "fringe"; second, a "margin" of data irrelevant to and so *not* covariant with the theme. As Gurwitsch says, "whereas variations of the thematic field ... entail modifications of the perspective under which the theme appears, no effect results from changes occurring in marginal consciousness" (*FC*, 415). So, in the example above (and, says Gurwitsch, in *any* field of consciousness), we have three domains of data:

(1) Theme: data at the focus of attention organized according to Gestalt law (e.g., thoughts of the First Crusade)
(2) Thematic field: data relevant to the theme and so covariant with it (e.g. related thoughts of romantic chivalry or holy wars)
(3) Marginal consciousness: data irrelevant to the theme and so not covariant with it (perception of desks, books, body, etc).

5. The margin of "mere co-presence" is itself structured, as Gurwitsch details in his posthumously published *Marginal Consciousness*. In particular, he claims that one's sense of body and world are always at least marginally present in the field of consciousness. Any marginal domain in any field of consciousness will contain at least this much: (1) data corresponding to bodily or "kinesthetic" awareness and (2) data corresponding to awareness of some sector of the physical world.[24] Even in the deepest historical rumination we have some sense of

[23] Some research in psychophysics argues against the possibility of a margin of irrelevant data. Baars (1988), on reviewing the empirical evidence, says "being conscious of any stimulus in any sense modality interferes with consciousness of any other stimulus, and also with conscious access to any voluntary act or conceptual process" (p. 100). I do not think this invalidates Gurwitsch's thinking; it rather suggests a graded account of degrees of relevancy.

[24] Gurwitsch actually parses marginal consciousness into three domains, the third being a "segment of the stream of consciousness," i.e. immediate memories and anticipations (in Husserl: retentions and protentions). These marginal data are of a qualitatively different

where we with our bodies are in the world. This is of course consistent with there being other data in the margin. Gurwitsch's point is that *at least* this much will always be present.

6. Gurwitsch describes a variety of "field transitions" whereby, in the course of ongoing experience, data in one domain of the field transition to another. In a thematic field → theme transition, we follow "lines of relevancy" from what is at the focus of attention to appurtenant ideas in the thematic field. I think now of the First Crusade, now of the concept of Chivalry, now of the Arthur legend. My thinking proceeds along Jamesean fringes from one thought to the next. On the other hand, a margin → theme transition is more abrupt, more jarring. The music that was in the background, irrelevant to your thoughts of the First Crusade, suddenly breaks through because it is so bad or because some ominous news breaks. Or perhaps a strange bird flies by. Or your name is called. In each case the train of thought from one relevant item to the next is broken by the intrusion of marginal data.

(B) Inter-Field Relations. Both fields are structured; neither is amorphous. The conscious field is structured into three domains of data that change over time in response to internal and external factors. The configuration of the brain's EM field also changes over time in response to internal and external factors. In this section I consider how the structure and dynamics of the conscious field might be related to the structure and dynamics of the brain's EM field and the neural activity which generates it. 1-6 following describe possible correlates of 1-6 above.

1. A basic assumption guiding these speculations is that "co-present data" in the field of consciousness correlate with spatial regions of the EM field generated by assemblies of spiking neurons. If there are 100 "data" in the field of consciousness corresponding to experienced objects in the surrounding world, one's sense of one's body, thoughts of the First Crusade, etc., then one might suppose there are 100 regions of the EM field correlated with these data.[25]

2. On Gurwitsch's account, an attentional "gestalt" or "theme" segregates itself from a non-focal (but itself structured) background. This suggests a similar partition of the EM field into regions associated with attention and other regions associated with background-consciousness (thematic field and margin). Some research has suggested that neural activity correlated with attention is character-

---

kind than the other two, and given the added complexities of time-consciousness, I leave this third domain of marginal consciousness out of consideration here.

[25] This relation need not be structure-preserving. Relevant data in consciousness may, for example, correlate with spatially distant regions of the EM field. Also note that no fixed relation between regions of the EM field and kinds of consciousness is implied, so that "activity in region *R* of the brain" would imply "experiential datum of class *C*." I make the minimal assumption that, if I no longer experience, say, a red rose, then there is some region of the EM field that must then have changed. On relations between "parts" of consciousness and spatial regions of the EM field, see Pockett, p. 108.

ized by increased synchrony and organization.[26] Let us call the regions of the EM field generated by this neural activity "attention-supporting regions" (as opposed to "non-focal" or "background-supporting" regions), leaving it open what exactly their unique characteristics are.

3. For Gurwitsch, recall, background or non-focal consciousness is itself structured into a thematic field of contextual data relevant to attention and a margin of data irrelevant to attention. Presumably these two kinds of background consciousness correspond to different non-focal regions of the EM field. Insofar as Gurwitsch distinguishes thematic field from margin on the basis of his concept of relevancy, this leads us to ask whether relevancy (a relation between data in the conscious field) correlates with some relation between spatial regions of the EM field which support relevant data. One possibility is that relevancy correlates with "coupledness" of the neural activity which generates these regions. In that case regions a and b of the EM field support *relevant* data in the field of consciousness when the activity patterns a' and b' which generate them are coupled.[27] Note that in speaking of "coupled regions of the EM field" in what follows I mean regions of the EM field generated by coupled assemblies of spiking neurons.

4. If the account of relevancy just given is right, then data in the thematic field (relevant to attention) are supported by regions of the EM field coupled with attention-supporting regions.[28] Data in marginal consciousness (*not* relevant to attention) on the other hand, are supported by regions of the EM field *not* coupled with the attention-supporting regions. So the three part structure of the conscious field into theme, thematic field, and margin correlates with a division of the electro-magnetic field into attention supporting regions, regions coupled with the attention supporting regions, and regions not coupled with the attention supporting regions. If this view is right, then my attentional musings on the First Crusade are supported by especially organized and synchronous regions of the

---

[26] Wolf Singer (2000) (who is referenced by field theorists but is not one himself) has argued that synchronous activity of "functionally coherent assemblies" of neurons supports the focal perception of Gestalts, while E.R John (p. 208) has proposed "negative entropy" as the critical quantity distinguishing figure from ground in an EM field.

[27] Populations of neurons (which generate regions of the EM field) are coupled if the state-equations which describe the rate of change of one population depend on the values of the state-variables other. Note that covariation applies to both relations. Relevancy of datum a to datum b in the conscious field implies that a change in a will affect the perspective under which b is experienced. Similarly with coupled regions of the EM field, where a variation in one region affects the activity in the other region. So in both cases—relevancy and coupling—change in one term entails (or better, increases the probability of) a change in the other.

[28] On the question of context-awareness and its functional significance, see Mangan (1993) and Baars (1997), who consider James' concept of "fringe" (which correlates with Gurwitsch's concept of "thematic field") vis-à-vis neurally-realized cognitive processes.

EM field, my context-awareness of Romantic Chivalry is supported by regions of the EM field coupled with this organized activity in the attention-supporting regions, and my marginal sense of body and the surrounding physical world are supported by regions of the EM field *not* coupled with the attention supporting region, but presumably generated by inputs from some other source.

5. Kinesthetic and perceptual inputs are constantly generating internally structured activity in the somato-sensory and perceptual cortex and make a constant contribution to the electro-magnetic field. This is consistent with the constant presence of bodily experience and perceptual awareness in marginal consciousness.

6. The transitional dynamics described by Gurwitsch may correspond to transformations in the EM field where background-supporting activity becomes attention supporting either in virtue of the connectedness of the underlying neural network or because of the influence of an external force. In the case of a theme → thematic field transition, where one moves along lines of relevancy in consciousness, the suggestion is that first one then another region of the EM field is made attention-supporting, based on propagation of electrical activity along paths in the underlying neural network. In the case of a margin → theme transition, on the other hand, where breaking news or an odd bird "break" into the focus of attention, the suggestion is that some region of the EM field which was *not* coupled with the attention-supporting region is made attention-supporting. Presumably this transition is initiated by some force in the external world corresponding to the unusual inputs associated with the news item or the bird.

Conclusion

Further work correlating Gurwitschean and neo-Köhlerian field theories is suggested by the above. For example, Gurwitsch culminates *The Field of Consciousness* with "ontological investigations" which expand his notion of relevancy. He describes "orders of existence": maximally expanded domains of relevance which are "pointed to" by the thematic field but only partially realized in experience (e.g., the order of physical objects, the order of mathematical objects, etc). It may be that orders of existence supervene on passive structures in the brain, in particular, on its passive but adjustable network of neurons. The intuition is that just as fields of consciousness are situated in largely unconscious (but explorable) orders of existence, so too are fields of electro-magnetic activity situated in webs of inactive (but activatable) nerve tissue.

But by now I have strayed from pure Gurwitschean thinking. He would allow these investigations of "mundane" conscious states and their causal correlates under the rubric of phenomenological psychology. But I suspect at places he would resist, perhaps for example in linking orders of existence with contingently structured webs of neurons in the brain. Of course at a deeper level these departures are themselves Gurwitschean. Again and again, Gurwitsch is an advocate of free and critical thought, faithful only to the phenomena. In the introduction to *Studies in Phenomenology and Psychology*, for example, he says: "no greater disservice could be rendered to [Husserl's, or in this case Gurwitsch's] work than 'freezing' it into a sectarian doctrine and jealously watching over the purity of the creed. Husserl never intended to found a sect. He initiated a working philosophy, a philosophy living and developing in the actual work of research" (*xxv*).[29]

References

Baars, Bernard. 1997. 'In the Theater of Consciousness: Global Workspace Theory, A Rigorous Theory of Consciousness.' *Journal of Consciousness Studies*, 4, pp. 292-309.

—. 1998. *A Cognitive Theory of Consciousness*. Cambridge, NY: Cambridge University Press.

Brook, Andrew. 2001. 'The Unity of Consciousness.' In *The Stanford Encyclopedia of Philosophy*. http://plato.stanford.edu/entries/consciousness-unity/

Carnap, Rudolf 1967. *The Logical Structure of the World*. Berkeley: University of California Press.

Embree, Lester. This volume. "The Three Species of Relevancy in Gurwitsch."

—. "Aron Gurwitsch's Theory of Cultural-Scientific Phenomenological Psychology." *Husserl Studies* 19:43-70, 2003.

Eliasmith, Chris. *Dictionary of Philosophy of Mind*. http://www.artsci.wustl.edu/~philos/MindDict/index.html

Fiset, P. Paus T, Daloze, T., Plourde, G., Muieret, P., Bonhomme, V., Hajj-Ali, N., Backman, S., and Evans, A. 'Brain Mechanisms of Propofol-Induced Loss of Consciousness in Humans: a Positron Emission Tomographic Study.' In *The Journal of Neuroscience*. 19 (13) 5506-5513.

Gurwitsch, Aron. 1929. 'Phenomenology of Thematics and the Pure Ego' in *Studies in Phenomenology and Psychology (SPP)*. Evanston, Ill: Northwestern University Press, 1966.

—. 1964. *The Field of Consciousness* (*FC*). Duquesne: Duquesne University Press.

—. 1966. *Studies in Phenomenology and Psychology* (*SPP)*. Evanston, Ill: Northwestern University Press, 1966.

—. 1974. *Phenomenology and the Theory of Science* (*PTS*). Evanston, Ill.: Northwestern University Press, 1974.

[29] I thank John Baressi, Jonathan Cohen, Lester Embree, Scott Hotton, Wayne Martin, Susan Pockett, Louis Sass, David Woodruff Smith, Olav Wiegand and especially Eric Thomson for helpful feedback and discussion.

—. 1985. *Marginal Consciousness* (*MC*). Athens, Ohio: Ohio University Press.

Hallett, M. 2000, 'Transcranial magnetic stimulation and the human brain,' *Nature*, 406, pp. 147–50.

Husserl, Edmund. 1900. *Logical Investigations*. Translation of the 2nd edition by J.N. Findlay. London: Routledge and Kegan Paul, 1973. [note: 1900 is the year the first ed. was published].

—. 1925. *Phenomenological Psychology. Lectures, Summer Semester 1925*. Translated by J. Scanlon. The Hauge: Nijhoff, 1975.

John E. R, L. S. Prichep, W. Kox, P. Valde′s-Sosa, J. Bosch-Bayard, E. Aubert, Tom, F. diMichele, and L. D. Guginoi. 2001. 'Invariant Reversible QEEG Effects of Anesthetics' *Consciousness and Cognition* 10, 165–183.

John, E. R. 2001. 'A field theory of consciousness.' *Consciousness and Cognition*, 10(2), 184–213.

Kim, Jaegwon 1993. *Supervenience and Mind.* Cambridge: Cambridge University Press.

—. 1998. *Mind in a Physical World.* Cambridge: MIT Press.

Köhler, Wolfgang. 1951. 'Relational determination in perception.' *Cerebral Mechanisms in Behavior: The Hixon Symposium.* ed. L. A. Jeffress, pp. 200-243. New York: John W.

—. 1959. 'The Mind-Body Problem', in Sidney Hook ed., *Dimensions of Mind,* Washington Square: New York University Press.

Köhler, Wolfgang and R. Held. 1949. 'The Cortical Correlate of Pattern Vision,' *Science* 110: 414-419.

Köhler, Wolfgang, R. Held, and D. N. O'Connell. 1952 'An investigation of cortical currents.' *Proc. Am. Philos. Soc.* 96:290-330.

Lashley, K. S., Chow, K. L., & Semmes, J. 1951. 'An examination of the electrical field theory of cerebral integration.' *Psychological Review*, 58. 123- 136.

Leopold, D.A., and N.K. Logothetis. 1999. 'Multistable phenomena: changing views in perception.' *Trends in Cognitive Sciences* 3: 254-264.

Mangan, Bruce. 1993. 'Taking Phenomenology Seriously: The 'fringe' and its implications for cognitive research.' *Consciousness and Cognition*, 2, pp. 89-108.

McLaughlin, Brian. 1995. 'Varieties of Supervenience.' in Savellos, Elias and Yalçin, Umir 1995. *Supervenience: New Essays*. Cambridge: Cambridge University Press.

McFadden Johnjoe. 2002a. 'Synchronous Firing and Its Influence on the Brain's Electromagnetic Field: Evidence for an Electromagnetic Field Theory of Consciousness.' *Journal of Consciousness Studies*, 9, No. 4, pp. 23–50

—. 2002b. "The Conscious Electromagnetic Information (CEMI) Field Theory." *Journal of Consciousness Studies*, 9, No. 8, pp. 45-60.

Polonsky Alan, Blake R., Braun J, and Heeger D. 'Neuronal activity in human primary visual cortex correlates with perception during binocular rivlalry.' *Nature Neuroscience* 3:11, pp. 1153-1159.

Pockett, S. 2000, *The Nature of Consciousness: A Hypothesis* (Lincoln, NE: Writers Club Press).

Ruohonen, J., Ollikainen, M., Nikouline, V., Virtanen, J. and Ilmoniemi, R.J. 2000, 'Coil design for real and sham transcranial magnetic stimulation,' *IEEE Trans Biomed Eng,* 47, pp. 145–8.

Savellos, Elias and Yalçin, Ümit. 1995. *Supervenience: New Essays*. Cambridge: Cambridge University Press.

Searle, John. 1983. *Intentionality: An Essay in The Philosophy of Mind.* Cambridge: Cambridge University Press.

Singer, Wolf. 2000. 'Phenomenal Awareness and Consciousness from a Neurobiological Perspective,' in *Neural Correlates of Consciousness*. Metzinger, Thomas ed. Cambridge: MIT Press.

Sperry, Roger. 1947. 'Cerebral regulation of motor coordination in monkeys following multiple transection of sensorimotor cortex,' in *Journal of Neurophysiology*. 10:275-94.

Sperry, R. W., Miner, N. and Myers, R. E. 1955. 'Visual pattern perception following subpial slicing and tantalum wire implantations in the visual cortex.' *J. Compar Physiol. Psych.* 48, 50-58.

Wiggins, Osborne. (this volume). 'Kinds of knowledge: Phenomenology and the Sciences.'

Yoshimi, Jeffrey. 2001. *Dynamics of Consciousness: Phenomenology, Neuroscience, and Dynamical Systems Theory*. Doctoral Dissertation, UC Irvine.

—. 'Mapping Supervenience.' Unpublished manuscript.

Chapter 7

The Marginal Body

Frédérique de Vignemont
New York University

Abstract

According to Gurwitsch, the body is at least in the margin of consciousness. If all components of the field of consciousness were experienced as equally salient, we would indeed not be able to think and behave appropriately. Though the body may become the focus of our conscious field when we are introspectively aware of it, it remains most of the time only in the background of consciousness. However, we may wonder if bodily states do really need to be conscious, even at the margin, or cannot be simply non-conscious. Action control requires permanent proprioceptive and visual feedback about the state and the position of our body parts. Experimental data shows that action monitoring operates at a non-conscious level and we may similarly suggest that we have a continuous unconscious access to bodily information. In this chapter, I thus intend to describe the various levels of body representations with the help of Gurwitsch's distinction. I will investigate the properties and the function of each of these levels.

Introduction

We spend the first year of our lives in self-exploration of our body. By looking at the mirror, playing with our limbs, touching our face and imitating others, we construct fine-tuned body representations that will allow us to recognize ourselves, to move, and to interact with the world (Rochat, 1998). After this early calibration, it seems that we no longer need to pay attention to our body, which has become a well-known tool that we frequently use without even thinking of it. The priority is given to the external world. Yet, this marginal body is not negligible for all that. Despite its peripheral status, it is still mentally represented, and even consciously. In this essay, I provide two compatible interpretations of the notion of marginal body: as bodily experiences in the background of consciousness and as unconscious sensorimotor schemata. With the help of Aron Gurwitsch's distinction (1964, 1985) and some clarifications of his main hypotheses, I intend to sketch the different levels of body representations.

L. Embree (ed.), Gurwitsch's Relevancy for Cognitive Science, 131-149.

I. The Visible and the Invisible Body

> This bodily presence is of a highly paradoxical nature. While in one sense the body is the most abiding and inescapable presence in our lives, it is also characterized by its absence. That is, one's own body is rarely the thematic object of experience. (Leder, 1990: 1)

How can the most permanent and preponderant object in our life be also the most elusive one? The presence-absence of the body can be understood at two different levels: the sensory one (which bodily information I receive) and the phenomenological one (what I feel).[1] At the sensory level, we continuously receive a flow of information about our own body through external and inner perceptions. Not only can we see or touch it, but we also have several inner receptors that convey sensory data about joint position, muscle stretching, pain, temperature, posture, balance, and nutrition. Unlike external perception, the inner sensory flow never stops and cannot be voluntarily controlled; you can close your eyes, but you cannot prevent the sense of balance. Thus, an important amount of information is always available, whether we want it or not, whether we pay attention to it or not. This "on-line" visual and proprioceptive feedback allows us to plan, initiate, and control our actions. At this basic level, the body is always present in all its details. However, it does not seem that this precision is preserved at the phenomenological level. I will start with a very general statement: except in a sensory deprivation state, we never feel disembodied and we can even feel as if a body part still exists when it has been amputated in the phenomenon of a phantom limb (Ramachandran and Hirstein, 1998). One's own body is characterized by its permanency and constitutes a point of reference for the dynamic world (Merleau-Ponty, 1945) and for the dynamic stream of consciousness (James, 1890). Each perceived object is encoded into an egocentric frame of reference centered on the body (Gurwitsch, 1985; Jeannerod, 1997). Moreover, according to William James, the bodily presence characterized by the "feeling of warmth and intimacy" would constitute the inner nucleus of the Self, that is, the intrinsic liaison between experiences. Thus, he suggested that the stream of thinking is in fact a stream of breathing. So, "the feeling of the same

[1] Similarly, Gallagher (2003) draws a distinction between proprioceptive information and proprioceptive awareness.

old body always there" (James, 1890: 242) would anchor both the external and the inner world.

Yet, most of the time this feeling is only in the background of consciousness. As Merleau-Ponty (1964) said, the body is "invisible": I do not see my eyes, I see the object seen by my eyes. It is true that I receive a large amount of information about the position of my eyes. Nonetheless, the focus of my attention is generally the external world rather than the medium that allows me to perceive it and to move through it. We use the body, but we rarely reflect upon it. In painful and learning situations, the body appears at the core of our interest, but when we walk in the street, we are never aware of the precise position of our legs and even less of our internal viscera state. This bodily absence could even explain Cartesian dualism (Leder, 1990): The prima facie fact is the *res cogitans* and not the *res extensa* of the body since the latter is less salient.

In summary, on the one hand, the body is not an object among others that could leave me or disappear; but on the other hand, even if all actions depend on it, it is rarely the focus of attention. Therefore, the question is to understand the cognitive organization that underlies this paradox pointed out by Leder. But is this really a paradox? We can avoid the contradiction if we assume that we are talking of body representations at different cognitive levels. The question is thus to determine which levels. Rather than a dichotomy between conscious and unconscious states, James and Gurwitsch introduced a distinction within the field of consciousness itself between core and fringe states. In this latter sense, we are always aware of our body. However, as we will see later, this awareness does not extend to all bodily states.

II. The Structure of Consciousness

William James and Aron Gurwitsch emphasized the permanence of the feeling of bodily presence, but acknowledged that the body usually stays in the background of consciousness. There is no paradox here because of the organization of consciousness itself. William James challenged the empiricist view according to which the field of consciousness should be conceived as chaos of experiences and claimed that conscious experiences are articulated into the focus and the margin. However, the organization stays extrinsic to the stream itself[2] and there is no distinction inside the margin: When I am reading a paper, the general knowledge that I have about its topic and my eye movements are put on the same level. Gurwitsch (1964) challenged this view and postulated an imma-

---

[2] Experiences are related by mutual compenetration and fringe states are integrated within each focus state, although varying in attentiveness. Thus, the organization is not intrinsic in the structure of consciousness and there is no real disconnection between the two kinds of states.

nent organization of the field of consciousness articulated by the intrinsic relation of relevancy between the material content of the theme and some co-present data. He claimed the existence of three (and not two) dimensions: the theme upon which the mental activities concentrate, the thematic field or context that is relevant to the theme, and the margin, which is co-present with the theme but irrelevant to it. Inside the margin, he also made a distinction between the halo that is externally related to the theme, for instance reflective thoughts, and the horizon that is only co-present with the theme. He used the metaphor of a circle. The theme constitutes the center of the circle, the thematic field is inside the circumference and the margin lies beyond the circumference. In summary, we can describe the structure of consciousness as below:

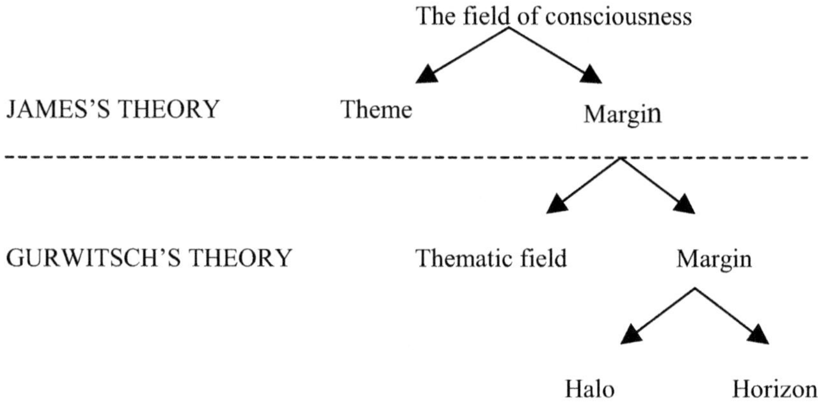

Fig. 1 The structure of consciousness

This schema does not exhaust the difference between James and Gurwitsch. The latter asserted that the margin is completely disconnected from the thematic field. In this sense, the marginal content of consciousness is contingent, because it does not need to respect any semantic constraint. This structure supposes that all the components of the field of consciousness are not experienced as equivalent, because otherwise we would not be able to order them and to choose one of them in order to behave appropriately. Three closely related factors intervene in the articulation of the field of consciousness:

- procedural factor: attention
- qualitative factor: intensity
- semantic factor: relevancy

These factors are not mutually derivable, although they may be related. Attention can be driven by the subject's deliberate decision (endogenous attention), but also by the saliency of stimuli, like their intensity (exogenous attention). For instance, a great pain will probably constitute the conscious theme, while if it hurts only weakly, it will stay at the margin. However, the most interesting feature of Gurwitsch's conception is the semantic factor.[3] Roughly, a state belongs to the thematic field if it is relevant to the theme, that is, if its content stands in a certain relationship with it. However, as we will see, the precise nature of this relation is not so easily defined.

In summary, marginal consciousness is contingent, irrelevant, and disconnected from the thematic field. We do not pay attention to it and it is dimly felt. Nevertheless, it allows us to always be aware of reality. At any moment, we can be conscious of three domains of reality, at least at the margin: (1) the stream of consciousness itself; (2) the perceptual environment; and (3) the bodily presence.

I will focus here only on this latter aspect. Even if we do not pay attention to our own body, it is still present at the margin. In this sense, there is no more paradox because the absence at the center of the circle can co-exist with a presence outside its circumference. The consequence of the described structure of consciousness is that we are always conscious of our body: "There is no moment in our conscious life when we are completely unaware of our bodily posture, of the fact that we are walking, standing, sitting, lying down." (Gurwitsch, 1985: 31)

Therefore, Gurwitsch's theory of body consciousness is articulated into two related assumptions: *The irrelevancy hypothesis* by which body consciousness is most of time marginal and irrelevant to the thematic field and *the conscious hypothesis* by which we are always conscious of our body.

Each of these hypotheses raises several questions. The first one implies that marginal body consciousness should never affect the theme. Thus, one should be as if she were disembodied at the level of the thematic activity. However, empirical results seem to argue against this disconnection: The body always interferes. The second hypothesis also requires some clarification. Gurwitsch did not describe the specific content of the marginal consciousness of the body and we can wonder about its nature and its precision. Once these questions are elucidated, we will be able to sketch the different levels of body consciousness and their relations.

---

[3] By semantic, I do not mean that we should understand the field of consciousness in linguistic terms or that all conscious states are propositional. Rather, I emphasize the importance of the content of the states, i.e., the intentional objects (or the noemata) of the states.

III. The Irrelevancy Hypothesis

According to Gurwitsch, marginal data are by definition irrelevant to the thematic field. He provided several illustrations and descriptions of the notion of relevancy, but as far as I know, he has never explicitly given any operational definition that allows one to decide when data are relevant for the theme.

> When we choose a geometrical figure as our theme, the mode of appearance of the figure varies according to whether the figure is presented on for example, a red or a yellow ground, whether the figure is surrounded by other figures, whether the surrounding figures or all of the same or of different kinds and so on. (Gurwitsch, 1964: 137)

> The passage [of music] under discussion refers to those preceding. It stands in definite and specific musical relations to those other passages and derives its musical perspective from such relation. (idem, 137)

> Absorbed in a scientific topic, e.g., a mathematical problem, we may think of a friend whose visit is anticipated during the course of the day. Disregarding the eventuality of our being distracted from our theme by the intruding thought, we have nothing more than two acts experienced simultaneously. If the anticipation of our friend's visit appears as a disturbing intrusion, this is because between the anticipated visit and the scientific topic engrossing our mind, merely the relation of simultaneous occurrence without any intrinsic relationship exists. (idem, 282-283)

> It may well be that, as James asserts, a closing of the glottis and an interruption of the breath are experienced, either regularly or occasionally, when we are confronted with a theoretical problem for which we do not see a solution … However, this bodily condition pertains in no sense whatever to the problem situation with which we are confronted and no change in our bodily condition affects the problem situation. No feature, tinge or aspect of the theme … derives from the actual bodily condition or is modified by an alteration of this condition. (Gurwitsch, 1985: 29)

> Besides being co-present with the theme, the data falling under the first class [the thematic field] appear, moreover, as *being of a certain concern to the theme*. They have something to do with it, they are relevant to it. Here the relationship is not merely that of simultaneity in phenomenal time, but is founded upon the material contents of both the theme and the co-present data. Such a relationship is intrinsic. (Gurwitsch, 1964: 340)[4]

What could we conclude? First, we need to understand that the mere phenomenal co-presence is not a sufficient condition of relevancy. Two simultaneous experiences are only related extrinsically by phenomenal time, whatever content they have. On the contrary, the relation of relevancy is intrinsic, based on the specific contents of the related states. Secondly, the relevant data interfere with the theme, modifying its visual mode of appearance or its musical perspective for instance. In other words, the theme would not be the same without the thematic field or with another thematic field. On the contrary, the marginal data do not affect the theme at all and seem to be completely disconnected from it.

Consequently, I suggest two possible conditions of the relation of relevancy. An experience A is relevant to the theme B: (i) If A belongs to the semantic network of B[5] and (ii) If A can affect the semantic content of B.[6]

For instance, if I am thinking about my family in France, then the memory of my last meeting with them seems to be relevant in both senses. On the contrary, the feeling that my legs are crossed appears as totally irrelevant. We can notice that the relation of relevancy is not reciprocal: A may affect B, without B affecting A. However, this definition of relevancy is far too wide. The notion of a semantic network is unclear. Roughly, the semantic network of B includes all the states that have the same class of meaning. More precisely, following Embree (2003), we can distinguish different species of relevancy: The perceptual relation between the figure and the background, the cultural relation between, for instance one utensil and its practical situation, the logical relation between different propositions inside an argument, and also the ontological relation within one domain of experiences like the inner life, the body, and the perceptual environment. However, these distinctions are not fine-grained enough for strictly delimiting the thematic field. In Gurwitsch's mathematics example, we cannot assume that mathematical knowledge is relevant: Geometrical principles

---

[4] Gurwitsch also qualified the relation of relevancy as a relation of "pertinence" (idem, 332), or of "affinity" (idem, 353), while irrelevant data were considered as "accessory, indifferent" (idem, 341).

[5] The thematic field is indeed defined as "being of a certain concern to the theme."

[6] The theme indeed "varies according" or "derives from" the relevant data and cannot be "modified" or "affected" by the marginal data.

are completely unrelated to the arithmetic problem that I have to solve. Therefore, the question is to determine how far we should extend the relevant semantic network of the theme.

The second condition may help. As Gurwitsch said, "no change in our bodily condition affects the problem situation." In other words, bodily consciousness should not affect the thematic activity. I will assume that the relevant data are able to affect the theme only if they can cause its modification in the same way as the background can alter the mode of appearance of the figure.[7] Still, the second condition does not describe what should be considered as an effect. Furthermore, it is controversial, as Natsoulas (1996, 1997) has pointed out. If I am attentively looking at an apple, then my body position affects my perception. He (1997) has emphasized this point by appealing to Gibson's conception of ecological vision: "Oneself and one's body exist along with the environment, they are co-perceived." (Gibson, 1979)

Natsoulas referred to Gibson's notion of visual kinesthesis: Each time I move, the patterns of the optical flow inform me that I am moving. Thus, the optical flow includes information about oneself as well as about the environment. However, this notion does not seem to be relevant in our discussion of Gurwitsch. Visual kinesthesis allows us to conclude that perceptual data are relevant to bodily data, but not the reverse. Nevertheless, Gibson also claimed that by perceiving my body, I perceive the world. Each perception is relative to one's own body position and state. I will provide three illustrations of this interaction between perception of the world and perception of the body.

First, the integration of touch with body position information is especially important for stereognosis, the use of touch to judge the size and the shape of objects.[8] Bermudez (1998) noticed that by touching an object in the night, the modification of the spatial properties of his fingers conveys new information about the spatial properties of the object. He concluded that proprioceptive consciousness is always at the periphery of tactile consciousness. Even if we do not pay attention to this proprioceptive consciousness, it still interferes with the tactile theme by allowing a better knowledge of the object.[9] Moreover, the misperception of the bodily position causes illusions in stereognosis. For instance, Aristotle discovered that by holding a small ball between the crossed third and

---

[7] We may notice that Gurwitsch was more interested in description than in the explanation. As Lester Embree points out in the introduction of this volume, causation is not the primary concern for Gurwitsch.

[8] Embree has also told in the introduction that Gurwitsch recognized the possibility of synaesthesis and we may then assume that he would accept the possible effect of bodily experiences in the perception of objects.

[9] However, I agree with Gallagher (2003) that it is most probably proprioceptive information rather than proprioceptive awareness that is involved in stereognosis. As we will see later, the unconscious status of the bodily information explains how it can affect the theme without being part of the thematic process.

fourth fingers, you feel as if the ball had doubled. This doubling occurs because the tactile input is interpreted as if the fingers were not crossed (Graziano and Botvinick, 2000).

Thus, the spatial organization of perceptual states is constructed by the subject through the interaction of visual, tactile, and kinesthetic inputs (Poincaré, 1905). For instance, vestibular information contributes to the conscious perception of orientation (Berthoz, 1997). Patients with lesions of the parieto-insular vestibular cortex—involved in the encoding of the movement of the head in space—see the world leaning to the side opposite to the lesion. The effect of deviations of the subjective vertical is so strong that their photographs are skewed (Dieterich and Brandt, 1993).

The neuroscience of vision emphasizes the importance of bodily information. We are not static in the world and the objects that we see often constitute goals for actions. Vision has two main functions: recognition and action. This functional distinction can be found at the anatomical level between the ventral visual pathway (infero-temporal) dedicated to the semantic process of the object and the dorsal visual pathway (infero-parietal) dedicated to the interaction of the subject with the object (Milner and Goodale, 1995). Vision for action computes visuo-motor transformations and encodes the relevant features for action into an egocentric frame of reference. These coordinates change through action execution and always need to be updated. For instance, the position in space of the apple that you want to reach is transferred from an extrinsic (visual) into an intrinsic (proprioceptive) system of coordinates (Jeannerod, 1997). The focus of your attention is the apple that you see, yet bodily information is necessary. Similarly, when you play tennis, you pay attention to the ball and not to your body (except if you are learning to play), but you will never be able to hit the ball if you do not compute its position in respect with your body.

We can provide two possible interpretations of these examples, which seem to challenge Gurwitsch's hypothesis of disconnection between the thematic field and the margin. We have just demonstrated that the representation of bodily properties makes a difference for the theme. Thus, we could conclude that in those cases, the consciousness of the body is not marginal, but belongs to the thematic field. This could be true in some cases but it cannot be true for all the phenomena described, because it would ultimately imply that most of the time the body is not at the margin of our consciousness, contrary to what Gurwitsch said. We are indeed always interacting with the world so that we always need to know the position and the state of our body. Except when these interactions are themselves marginally conscious, the body would always belong to the thematic field. In order to avoid such conclusion, we then may suggest that bodily consciousness stays at the margin because even if it fulfills the second condition, it does not satisfy the first one. For instance, my desire to eat the apple in front of me is not semantically related to the proprioceptive experience of my arm moving. As Natsoulas (1997) said, even if body information affects the theme, it is not included in the thematic process. Similarly, Gurwitsch (1985: 29) concluded that bodily condition is not "integrated into the thematic process."

We have just tried to investigate more deeply the notion of relevancy in order to understand how bodily consciousness can be marginal, despite the almost constant interference of bodily information with the theme. The combination between the first condition of the semantic relation and the second condition of the effect relation seems to be able to restrain the extension of the notion of relevancy. However, as Gurwitsch (1964) himself acknowledged, the frontier of the thematic field remains indeterminate:

> Not only does the thematic field admit indistinctness and indeterminateness, but, as previously mentioned, vagueness, and indetermination usually prevail in the more remote zones of the thematic field. A given theme points and refers not only to the 'adjoining' parts of the thematic field, that is, objects, facts, data, and items of immediate concern to it, but also to more remote zones having comparatively little bearing upon the theme, and even beyond to items fusing with one another into an inarticulate mass. Such a mass has hardly a direct, if any, concern to the theme. Still, that fused and confused mass is referred to as somehow relevant to the theme, at least in that its indistinct and indiscriminate components having "something to do" with the theme or are of the same kind and nature ... Properly speaking, the experience of relevancy has here the sense of *indefinite continuation* rather than *continuity of context*. (Gurwitsch, 1964: 378-380)

Thus, we can still wonder whether the bodily data do not belong to the most remote zone of the thematic field. One way to avoid this conclusion is to suggest that they do not even belong to the field of consciousness. In the next part, we will see the conditions under which body information can have "something to do" with the theme without being part of the thematic field.

IV. The Consciousness Hypothesis

> Our own bodily position, attitude, condition, is one of the things of which some awareness, however inattentive, invariably accompanies the knowledge of whatever else we know. We think, and as we think, we feel our bodily selves as the seat of our thinking. (James, 1890: 242)

> We are immediately and directly aware of our body, at least in marginal form, at every moment of our lives, under all circumstances, and at whatever place we might happen to find ourselves. Again, the body is the only individual mundane object which is thus permanently present. (Gurwitsch, 1985: 60)

James and Gurwitsch claimed that we are always aware of our body. I do not want to challenge this view, but rather to investigate the content of the continuous marginal body consciousness. Do I feel the presence of my body ("I have a body") or its state ("I am sitting")? It is right that we feel embodied, except in some depersonalization disorders where people have a sensation of dissociation from their own body. For instance, they may feel as if they were dead (Simeon et al., 2000) or floating above their own body (Grotstein, 1983). Descartes also doubted the reality of his own body; nothing assured him that he had a body, that he was not dreaming or hallucinating its existence. More recently, Putnam (1981) has suggested that we could be disembodied brains in a vat and that the experience of our own body would be the result of electric impulses from a computer. Nevertheless, neither Descartes nor Putnam refuted the fact that we have bodily feelings, they only tried to challenge the fact that there is any objective body beyond these feelings and even in this respect, they both refuted skepticism.

However, Gurwitsch went further and claimed that we are not only aware that we have a body, we are also aware of its position and its state. It does not mean that the content of body consciousness at the margin is as detailed as it would be if it were the focus of our attention. On the contrary, the marginal awareness is "dim, vague, indistinct, and indeterminate" (Gurwitsch, 1985: 27) and constitutes "merely the indiscriminate and obscure horizon, devoid of inner structure and articulation" (idem, 35). The bodily experiences are only co-present at the margin and they can be integrated into a consistent representation of the body only at the level of the thematic field. However, are we conscious of the body in all its details or only as a unitary whole without any internal differentiation?

O'Shaughnessy (1980, 1995) raised the same question. In 1980, he argued that we should be conscious of all the points of the body at any moment. He noticed that we are able to realize whatever action whenever we want without having to think about the position of our limbs. This continuous flow of information that we use should therefore always stay at the margin of consciousness. However, he acknowledged in 1995 that his theory was not parsimonious and did not fit well with reality. We consciously perceive our body as an undifferentiated whole and actions automatically trigger a precise perception of the limbs involved, while the other parts of the body recede in the awareness. Thus, proprioceptive consciousness would be similar to peripheral visual perception (Kinsbourne, 1995): Visual information at the periphery of the retina is less specific, as there are fewer photoreceptors; similarly, the content of marginal proprioceptive representations is less fine-grained (I am conscious that my arm is flexed, but not of its precise angle).

Several experimental data show that we are weakly conscious of our body position, even at the margin. Most of our actions are automatic and we have a limited awareness of motor and body representations. For instance, Fourneret and Jeannerod (1998) asked subjects to trace sagittal lines on a graphic tablet while they were given visual feedback projected from a computer screen on a mirror. In normal trials, the line seen in the mirror exactly matched the traced lines. In perturbed trials, a bias was introduced and subjects consistently displaced their hand in order to compensate the bias. After each trial, they were asked in which direction they thought their hand had moved. They showed poor consciousness of the signals generated by their own movements and they tended to adhere to visual rather than to proprioceptive representations. For instance, they reported the opposite direction of their actual movements (e.g., a bias of 15° to the left rather than to the right). Therefore, most of the time, we are conscious only of the general goal of the action and the visual feedback of the execution, but the specific parameters of the way we have accomplished the movement—that is, the succession of body positions—are unavailable to awareness, even at the margin. We need thus to make a distinction between, on the one hand, the body representations used for action control that are fine-grained, but that are not available to consciousness and, on the other hand, the body representations that are conscious, but not necessarily accurate, nor detailed.[10]

Gurwitsch provided a conceptual framework for understanding the status of the body in respect to consciousness. However, we have just seen that we also need to take into account the status of the body outside the sphere of consciousness. The metaphor of the circle is no longer appropriate because it implies that there is nothing beyond marginal consciousness, as it seems to include all that lies outside the circumference. Thus, there seems to be no room for unconscious bodily representations. Nevertheless, the experimental data show that we need to postulate another dimension beyond the circle, that is, unconscious body representations. We should not assimilate perception, attention, and consciousness. The body is represented on the basis of the continuous flow of sensory information, prior to any thematization in the conscious field. Even at the unconscious level, we consider the body as being mentally represented.[11]

---

[10] If action execution used only conscious body representations, it would not be reliable because they are not fine-grained enough to allow us to reach any object.

[11] The unconscious states cannot be reduced to a flow of neural activity. They are properly mental and they do far more than simply pool the sensory information. Rather, they integrate the sensory inputs from different modalities into a meaningful representation by establishing new relationships between them.

The question is then to understand why bodily representations sometimes become conscious. It raises the problem of the relation between attention and consciousness: Is attention a necessary condition for consciousness or can we only pay attention to what is already conscious? This question is all the more difficult that psychological literature often confuses both notions. Moreover, it goes far beyond the body domain. Gurwitsch focused on the attentional process that occurs in the field of consciousness itself. However, we should not preclude the prior existence of automatic attentional selectivity. Attention intervenes at different stages in the process of information, and more particularly, we should distinguish between the voluntary and selective attention and the low-level process of the attentional filter. Various experimental results show that attention may even be a prerequisite of consciousness (Dehaene and Naccache, 2001). Conditions of stimulation by themselves, such as duration and clarity, do not suffice to elicit consciousness. For instance, people with right parietal lesions suffering from personal neglect explicitly deny the existence of the contralateral side of their body or forget it (e.g., they shave or make up only the right side of their face). Nevertheless, if you allow them to pay attention to the forgotten side, then they become conscious of it: When the two hands are crossed, such that the left hand was to the right of the trunk and conversely, the tactile extinction switches hands (Aglioti, Smania and Peru, 1999). Moreover, objects that do not fall in an attended region of the visual field cannot be consciously reported, as it is shown by the phenomenon of inattentional blindness (Mack and Rock, 1998). The subject is required to perform a visual discrimination task at a specific location of his visual field, while another visual stimulus appears at different locations. The conditions of stimulation should suffice to elicit consciousness, and yet a large percentage of subjects failed to report the presence of the rival stimulus. In other words, body representations cannot be conscious if the subject does not pay any attention to them.

Therefore, we need to improve the metaphor of the circle in order to integrate unconscious representations of the bodily and at least two levels of attention (See Figure 2).

We can now reply to the problem raised by the irrelevancy hypothesis, that is, to the fact that some body information affects the thematic consciousness. The role played by body information in action monitoring does not imply that representations of bodily position are included in the thematic process, because they are not available to awareness, marginally or not. Actually, as we have already said, the content of marginal body consciousness is too coarse-grained for action monitoring. Therefore, we can maintain the disconnection hypothesis between thematic and marginal consciousness. The nature of unconscious representations of the body and their relations to the thematic process are another question.

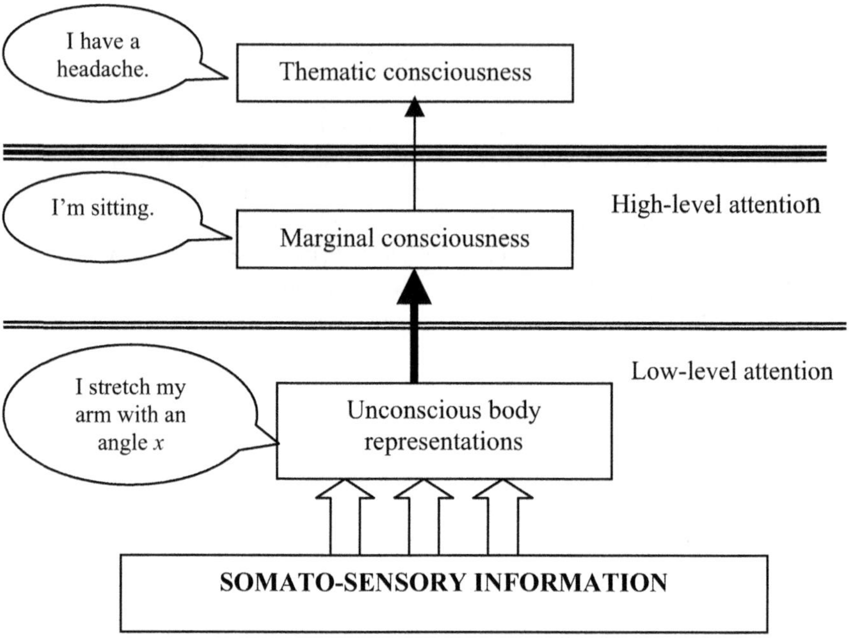

Fig. 2 Perception, Attention and Consciousness

V. Body Schema and Body Image

Let us imagine that I am attending to a very interesting conference about Gurwitsch. When I enter into the room, I have a brief look at myself in the mirror: I consciously represent my body, mainly based on visual information, I think that I have shadows under my eyes and feel sorry about it. Then, I take a seat and listen carefully to the philosopher who is talking. Yet, I dimly experience that I have a body, which is complete and which belongs to me, and that I am sitting. During the talk, my pen falls and I bend down in order to grasp it. The sensori-motor system encodes the position of the pen relative to my body and monitors my movements in order to catch it.

Thus, far from being marginal the body constantly intervenes in our life involving different kinds of representation that vary in respect with several dimensions. We may summarize this example as below: (See Fig. 3)

	Consciousness	Specification	Dynamics
Looking at myself in the mirror	Thematically conscious	More or less detailed	Long-term and short-term representations
Focusing on the topic of the conference	Marginally conscious	Body as a whole	Long-term and short-term representations
Grasping the pen	Unconscious	Very accurate	Short-term representations

Fig. 3 Reflexive body, Marginal body and Unconscious body

Head and Holmes (1911) suggested the existence of three kinds of body representation: the *body schema*, which reflects the modifications of position that are not yet conscious; the *superficial schema*, which is a central map of somatotopic tactile information; and the *body image*, which is a plurimodal conscious representation of the body. I thus suggest looking for the function of each of these representations.

As we have already seen, Milner and Goodale (1995) distinguished vision for recognition ("What") and vision for action ("How"). Similarly, we can distinguish between the body schema and the body image (Gallagher, 1995). This functional distinction could also be applied to the representations of the body. Some neuropsychological results indeed show a double dissociation between patients with "numbsense" who cannot identify their body parts while they are still able to point them with their hand and deafferented patients who recognize them without being able to reach them (Paillard, 1999). Therefore, body representations differ with respect to the semantic or pragmatic nature of their purpose. In this sense, the body image replies to the question "What" (What is this body part? What is the position and the state of my body?), while the body schema replies to the question "How" (How to use my body?). The body schema allows us to walk, to grasp a glass or to catch a ball, while the body image is oriented toward the semantic recognition of the body.[12] From this functional difference, I will briefly describe several distinct features of these representations.

The body schema is a dynamic sensori-motor representation based on the continuous flow of somesthesic and visual information. If you want to move, you indeed need to know very quickly the position of your limbs at every moment. According to Libet et al. (1991), only cerebral activity that lasts more than 500 ms is able to elicit awareness. Therefore, as a short-term representation involved in action monitoring, the body schema is not available to consciousness.

[12] It may play a role for action, but only when it requires having a conscious and reflective behavior, like in motor training (Gallagher and Cole, 1995).

On the contrary, the body image involves long-term conscious representations that includes perceptual, conceptual, and emotional components. It relies on different conceptions of the body:[13]

- "the innate body": innate internal model of the body at the origin of the phenomenon of phantom limb in aplasic patients (Melzack, 1990)
- "the usual body": representation of the body stored in long-term memory
- "the actual body": visuo-spatial representation of the body
- "the semantic body": conceptual and linguistic representation of the body
- "the cultural body": social representation of the body and its evaluation by a specific culture

We can become aware of each of these body images. However, when they are only at the margin of the field of consciousness, they are less organized and articulated and we can have access to full-content body images only if we introspectively pay attention to our body.

Despite their differences, the body image and the body schema are intimately related. The body image does indeed structure the body schema, while it is itself partly the result of the representation common to the successive body schemata (O'Shaughnessy, 1995). However, it does not follow that the body image is only the part of the body schema that becomes available to awareness through the filter of attention. Their content and their functions are clearly distinct and should not be confused.

Conclusion

The body can be apprehended both as a point of view on the world and as an object within this world. The thematic consciousness, based on perceptual, conceptual, and emotional body images, emphasizes this latter aspect. However, we scarcely focus on our body because we are more involved in the external world than with ourselves. Nevertheless, we cannot have any interaction with mundane objects if we do not take into account the body itself. The implication of the body does not need to be conscious, and even our efficiency depends on the unconsciousness of sensori-motor body representations. Thus, the body schema expresses the body as an agent in its continuous relation with the world. Yet, the question of body representations cannot be settled by this dichotomization between conscious and unconscious processes, between object and subject. In-

---

[13] The distinction between the usual and the actual body is made by Merleau-Ponty (1945).

between these two levels, the marginal body consciousness plays its own role. Like the body schema, it is continuously present, always in the background of consciousness. Like the body image, it represents the body as a whole and does not intervene in action monitoring. Marginal bodily consciousness manifests the feeling that my point of view on the world is always anchored in this body that belongs to me.

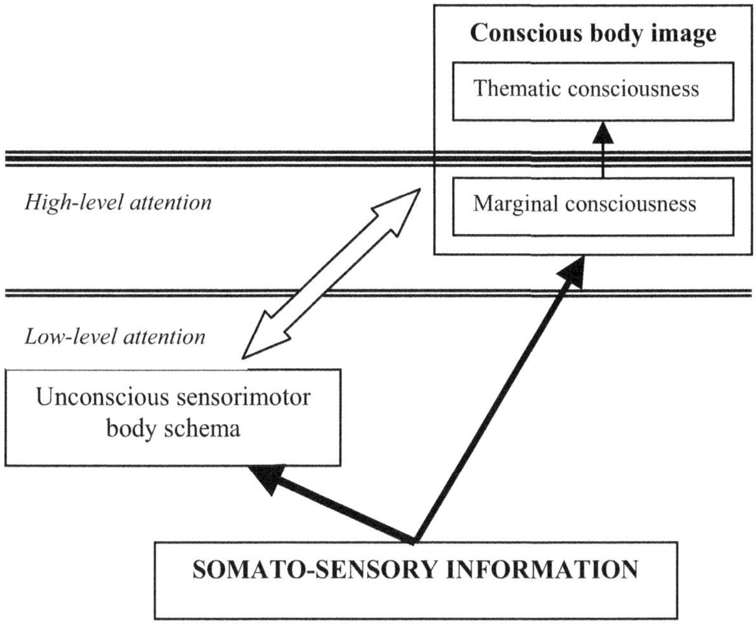

Fig. 4 Body representations

Acknowledgements

I am grateful to Lester Embree for the wonderful organization of the conference, "Gurwitsch's Relevancy for Cognitive Science" (Florida, 2002), and for his helpful comments, as well as to all the members of the conference for their questions.

References

Aglioti, S., N. Smania, and A. Peru. 1999. Frames of reference for mapping tactile stimuli in brain-damaged patients. *Journal of Cognitive Neuroscience*, 11: 67-79.

Berger, H. and F. Gerstenbrand. 1981. Phantom illusions in spinal cord lesions. In *Phantom and Stump Pain*, edited by J. Siegfried and M. Zimmermann, Springler-Verlag.

Bermudez, J.L., A. Marcel, and N. Eilan. (eds) 1995. *The Body and the Self.* Cambridge (Mass.): MIT Press.

Bermudez, J.L. 1998. *The Paradox of Self-Consciousness*. Cambridge: MIT Press.

Berthoz, A. 1997. *Le sens du mouvement.* Paris: Odile Jacob.

Dehaene, S. and L. Naccache. 2001. Toward a cognitive neuroscience of consciousness: basic evidence and a workspace framework. *Cognition*, 79: 1-37.

Descartes, R. 1641/1970. *Méditations Métaphysiques*. Paris: PUF.

Dieterich, M. and T. Brandt. 1993. Thalamic infractions: differential effects on vestibular function in the roll plane. *Neurology*, 43: 1732-1740.

Embree, L. 2003. The three species of Relevancy in Gurwitsch. This volume.

Fourneret, P. and M. Jeannerod. 1998. Limited conscious monitoring of motor performance in normal subjects. *Neuropsychologia*, 36 (11), 1133-1140.

Gallagher, S. 1995. Body schema and intentionality. In *The body and the self*, edited by J.L. Bermudez, A. Marcel and N. Eilan, Cambridge: MIT Press.

—. 2003. Bodily self-awareness and object perception. *Theoria et Historia Scientiarum: International Journal for Interdisciplinary Studies*, 7(1) – in press

Gallagher, S. and J. Cole. 1995. Body Schema and Body Image in a Deafferented Subject. *Journal of Mind and Behavior*, 16 (4), 369-390.

Gibson, J.J. (1979), *The Ecological Approach to Visual Perception*. Boston: Houghton Mifflin.

Graziano, M.S.A. and M. M. Botvinick. 2000. How the brain represents the body: insights from neurophysiology and psychology. In *Attention and Performance*, edited by Prinz

Grotstein, J.S. 1983. Autoscopy: the experience of oneself as a double. *Hillside J. Clin Psychiatry*, 5(2): 259-304.

Gurwitsch, A. 1964. *The Field of Consciousness*. USA: Duquesne University Press.

—. 1985. *Marginal Consciousness*. Athens: Ohio University Press.

—. A. 2002. *Esquisse de la phénoménologie constitutive*. Paris: Vrin.

Head, H. and G. Holmes. 1911. Sensory disturbances from cerebral lesions. *Brain* 34, 102-254.

James, W. 1890. *The Principles of Psychology*, Vol 1. New York: Holt.

Jeannerod, M. 1997. *The Cognitive Neuroscience of Action*. Oxford: Blackwell.

Kinsbourne, M. 1995. Awareness of One's Own Body: An Attentionnal Theory of Its Nature, Development, and Brain Basis. In *The Body and the Self*, edited by J.L. Bermudez, A. Marcel, N. Eilan, Cambridge (Mass.): MIT Press.

Leder, D. 1990. *The Absent Body*. Chicago: University of Chicago Press.

Libet, B., D. K. Pearl, D. E. Morledge, C. A. Gleason, Y. Hosobuchi, N. M. Barbaro. 1991. Control of the transition from sensory detection to sensory awareness in man by the duration of a thalamic stimulus. The cerebral 'time-on' factor. *Brain*, 114, 1731-57.

Mack, A. and I. Rock. 1998. *Inattentional Blindness*. Cambridge (Mass.): MIT Press.

Melzack, R. 1990. Phantom limbs and the concept of a neuromatrix. *Trends in Neuroscience*, 13, 88-92.

Merleau-Ponty, M. 1945. *Phénoménologie de la perception*. Paris: Gallimard.

—. 1964. Le *visible et l'invisible*. Paris: Gallimard.

Natsoulas, T. 1996. The stream of consciousness: XII. Consciousness and self-awareness. *Imagination, Cognition and Personality*, 16(2): 161-180.

—. 1997a. The stream of consciousness: XIII. Bodily self-awareness and Aron Gurwitsch's margin. *Imagination, Cognition and Personality*, 16(3): 281-300.

—. 1997b. The stream of consciousness: XIV. Two contrasting accounts of pervasive bodily self-awareness. *Imagination, Cognition and Personality*, 17(1): 45-64.

O'Shaughnessy, B. 1980. *The Will: Dual Aspect Theory*. Cambridge: Cambridge University Press.

—. 1995. Proprioception and the body image. In *The Body and the Self*, edited by J.L. Bermudez, A. Marcel and N. Eilan, Cambridge (Mass.): MIT Press.

Paillard, J. 1999. Body schema and body image—A double dissoiation in deafferented patients. In *Motor Control, Today and Tomorrow*, edited by G.N. Gantchev, S. Mori and J. Massion. Sophia: Academic publishing House.

Poincaré, H. 1905. *La science et l'hypothèse*. Paris: Flammarion.

Putnam, H. 1981. *Reason, Truth and History*. Cambridge: Cambridge University Press.

Ramachandran, V.S. and W. Hirstein. 1998. The perception of phantom limbs. *Brain*, 121: 1603-1630.

Rochat, P. 1998. Self-perception in infancy. *Experimental Brain Research*, 123: 102-109.

Chapter 8

Experimental Evidence for Three Dimensions of Attention

P. Sven Arvidson
Seattle University

Abstract

Using recent cognitive scientific research and Aron Gurwitsch's insights, I argue that the structure of attention is best thought of as tripartite. There are three distinct organizational principles coordinate with three dimensions of attention, and the focus of attention is only one of these three dimensions. Attention is a unified, dynamic embodied processing in the world, but can be seen phenomenologically and experimentally to operate in three distinct dimensions or domains. The theoretical context for attention experiments has always defined attention in terms of the focus or target. Gurwitsch's writings also define attention as constrained to this one dimension, the theme. Yet allowing also for contextual attention and marginal attention squares well with recent experimental findings and with the implicit thrust of Gurwitsch's phenomenology. With this expanded definition and function for attention, I interpret Gurwitsch's thematic field and margin dimensions of consciousness as dimensions of attention.

Introduction

No systematic attempt has ever been made to marshal experimental evidence for Aron Gurwitsch's key claim that there is always a theme, thematic field, and margin in human conscious life. This essay is just such an attempt. The result: the work and findings of cognitive scientists experimenting on attention essentially substantiates Gurwitsch's phenomenological claim. In order to line up phenomenology and cognitive science so that this collaboration can be made, I make adjustments on each side. On Gurwitsch's side, I interpret his three dimensions of the field of consciousness as correlated with three types of attention processing. On the side of cognitive science, I interpret processing outside the focus as different types of attention. I believe these interpretations preserve the essential findings of each side and afford mutual confirmation and enlightenment previously not available. I also recognize that it would be easy to find members on each side who would strenuously object to my moves here.

L. Embree (ed.), Gurwitsch's Relevancy for Cognitive Science, 151-168.

However, I believe these interpretations are not only warranted but extremely useful, as explained below. As in my related work, this essay seeks to push both Gurwitsch's philosophy and cognitive scientific paradigms of attention beyond their current conceptions (Arvidson, 1996, 1997, 1998, 2000, 2003a, 2003b).

Before proceeding, I will briefly elaborate five important points. First, for Gurwitsch and cognitive scientists "attention" refers exclusively to the achievement of the target, focus, or theme in the field of presentation, while my use is much wider. Certain anomalies and odd locutions in current research on attention, such as "unattended stimuli" (for stimuli that nevertheless appear to be somehow attended to) raise the issue of how that which is outside of the focus of attention is attended to. My response is to regard the field of consciousness as an attentional field, and to divide it into three dimensions, each with its own organizational principle. That is, there are three distinct organizational principles coordinate with three dimensions of attention, and the focus of attention is only one of these three dimensions. Attention is a *unified*, dynamic embodied processing in the world, but can be seen phenomenologically and experimentally to operate in three distinct dimensions or domains.

This leads to the second important preliminary point. This wider sense of attention means that I would use the word "attention" where Gurwitsch would use "consciousness" or "awareness"—namely, in referring to the thematic field or the margin. As this essay shows, there is strong evidence for thinking of the processing of content in the thematic field and in the margin as different types or modes of attention, rather than as modes of consciousness. With this expanded sense of attention, there is no need to use the vague, less operationally definable word "consciousness."

Third, use of the concept of "dimension" does not imply a static model of attention. In fact, Gurwitsch was among the first to categorize radical transformations of attention and to even discuss the notion of a transformation law that structures attention as a process. Following what he implies, I take attention to be *a dynamic tension of transformations of content involving dimensional organizational principles and transformation principles.* Attention is a process, it is dynamic and often tense; it can involve significant transformations of content and relationships. This means that the standard of measurement in cognitive science research on attention—the stable achievement of a target—is but a small part of how attention actually works. As a whole, it is much more dynamic and exciting.

Fourth, I draw my inspiration from Gurwitsch's work and see my own implied in his. But there may be plenty of reason for Gurwitsch to have been critical of what is presented here.

Fifth and finally, one might object to the sense of this project and claim that the analysis of attention here is merely a substitution of the word attention for the word consciousness, as the word is used for example in "thematic field consciousness" or "marginal consciousness," without any other significant implications.

Yet once one moves beyond Gurwitsch's view that attention is only a targeted or focused conscious process, a thematic consciousness, then a number of positive findings result. I will name six that I think are most important. (1) Most generally, a parallelism is established between current cognitive scientific research, including neurological discoveries, and Gurwitsch's philosophy—a mutually illuminating parallelism such that advances on one side may translate into advances in the other. (2) Gurwitsch's philosophy of consciousness (which includes at its core a philosophy of attention) becomes a philosophy of attention, and so the bifurcation between attention and consciousness disappears, and the terminological and systematic unity of his descriptions of human experience are enhanced. (3) Gurwitsch's philosophy can help the interpretation of experimental data, set research agendas, and define experimental paradigms in cognitive science, a possibility of intimate association between current research in the laboratory and phenomenology that is rare in today's Balkanized scientific practice/philosophy of science milieu. (4) Gurwitsch's very important transformation principles, which are principles of attentional modification that involve more than just the theme, can now become the focus of hypothesis and experimentation by cognitive scientists working on attention precisely because attention is no longer limited to the focus or theme. (5) The nature of "thematic field consciousness" and "marginal consciousness," what they are and what happens in those domains, is opened up to a systematic analysis and taxonomical discovery that only a two-pronged investigation from both phenomenology and cognitive science can effect, so that the regions, modifications, and conditions of these domains and processes are unveiled in a way that neither approach could accomplish alone without the recognition of these domains as attentional rather than conscious. (6) Identifying "marginal consciousness" as "marginal attention" affords a better account of Gurwitsch's concept of non-egological subjectivity.

Even if one rejects the identification of consciousness with attention in three dimensions, the crucial point of this paper is not subverted. Namely, *this essay shows that Gurwitsch's central claim that all content is structured in three dimensions in the field of consciousness at every moment of conscious life is supported by current cognitive science research.* Now to the evidence.

I. Theme, Thematic Context, and Margin

Inspired by the work of William James and Edmund Husserl on the invariant regularities of how things are presented in consciousness, Gurwitsch formulates a three-part field of consciousness—theme, thematic context, and margin (I substitute the term "thematic context" for Gurwitsch's "thematic field"). He claims that all data or content presented in consciousness belongs to one of these three dimensions. He (1966, pp. 267-8) writes that the total field of consciousness can be symbolized by a circle: "The theme with which we are dealing occupies the center of this circle; it stands in the thematic field, which—to abide by the metaphor—forms the area of the circle; and around the thematic field, at the periphery as it were, the objects of marginal consciousness are arranged." For the purposes of this essay, "figure," "focus," and "theme" are taken as roughly equivalent terms, as are "ground," "context," and "thematic field." Taken together, this section and the next make the point that *all three* of these dimensions of attention are being used or referred to implicitly or explicitly in research on attention, and hence exist as Gurwitsch claims they do.

A key concept for Gurwitsch and phenomenology is *intentionality*.[1] When describing conscious organization, Gurwitsch distinguishes between the field of consciousness (all that is present*ed* or intend*ed* in consciousness) and the consciousness of field (the present*ing* or intend*ing* activity). Outside of the phenomenological frame, this distinction refers to that between the contents and acts of consciousness. Intentionality names the correlation between the field of consciousness (noemata) and the consciousness of field (noeses). For Gurwitsch, investigation of the field of consciousness involves articulating the organization of what is presented in consciousness *as* it is presented. As Martin Heidegger (1962, p. 58) puts it: "Thus 'phenomenology' means…to let that which shows itself be seen from itself in the very way in which it shows itself from itself." Although they are correlated, continuity in the field of consciousness is articulated in terms of attentional transformations (Arvidson, 2000), and continuity in the consciousness of field in terms of temporal transformations (Husserl, 1991).

Any content can appear in any of these three domains. The fresh coffee spill on my desk engrosses my attention, it is presented as thematic; it stands out from but is presented within a thematic context that is relevant to it. This thematic context is the theme's background, namely, the desk upon which it lays, its position near the phone, etc. Both the theme and the thematic context content are distinct from the marginal content of my itchy allergy nose. However, as the itchy nose captures my focal attention, which is to say, as it vies for and wins thematic status, the field of attention is transformed significantly. My allergy pill bottle (perhaps in the next room?) becomes relevant to the new

[1] Contrary to the views of Gurwitsch and Edmund Husserl, I have suggested elsewhere (Arvidson, 2003a) that intentionality reduces to attentionality (cf. Husserl, 1982,§92).

theme of my itchy nose; this means that the bottle is attended to as part of the thematic field or context for the itchy nose as theme. Also, the previous contents of the theme and thematic context are now attended to as marginal, the coffee ring and the desk, phone, etc., are paid attention to differently than before. Is the pill bottle in the next room? Where is it? Now the pill bottle is thematic. I have simplified this example tremendously in order to show, in a preliminary way, both the peculiar nature of the three dimensions and the possibility of a transformation between them. Attention is much more complicated than my description suggests. For example, in the last part the pill bottle is thematically presented *as absent*, and in the first part the itchy nose that first appears as marginal is a "distractor" that eventually causes interference with attention to the "target" (the fresh coffee spill) and supplants it in "attention capture."

Any recent cognitive science experiment on attention, including neurological approaches, assumes the existence of a focus or theme of attention and then proceeds to measure it in some way. I would agree with this assumption, as would Gurwitsch. However, I cannot follow Gurwitsch and cognitive scientists when they implicitly or explicitly limit attention to the focus, target, or theme. In what follows, I will show how there is contextual attending and marginal attending. Each is a distinct *type* of attending, but each is present at every moment, along with thematic attending. *Attending thematically* (i.e., focally) allows content to become consolidated as a Gestalt, segregated from the thematic context, and *centralized* in it. *Attending contextually* (thematic context attending) allows content to become consolidated as non-centralized Gestalts, *relevant* to the theme and to other thematic context content (rather than being thematic themselves). *Attending marginally* allows content to become consolidated as non-centralized Gestalts, segregated from the theme and thematic context, and co-present but *irrelevant* to them. By "contextual" and "marginal" attending I mean nothing more than what Gurwitsch means by thematic field consciousness and marginal consciousness. The difference is the framework for interpretation of the phenomenological and cognitive scientific findings, namely, a framework of attention processing.

All agree that thematic attention aims to accomplish a single, centralized Gestalt consolidated and segregated from the rest of the field. The content of this theme can be made up of numerous constituents, such as a crowd of people or an orchard of apple trees, but the aim is still a centralized unity demarcated from the rest of what is presented. Gurwitsch makes the further claim (Gurwitsch, 1964, pp. 103, 336n), which I have amplified (Arvidson, 2000) that this achievement can still be called attention even if the theme is partially formed, dim, fleeting, etc. The difference between attending to something focally and attending to something contextually (that is, the difference between theme and thematic context attending), is not the presentation of Gestalts as consolidated content. The difference is that attending to something contextually means allowing this consolidated content (Gestalt) to be organized as relevant to the theme, and as relevant to other content. Attending contextually brings about a "unity by relevancy" for certain co-present content (the thematic context), based upon the intrinsic material relatedness of the theme and this content, such that the thematic context is a "domain of relevancy" with the theme as its central point of reference (Gurwitsch, 1964, pp. 340-342; p. 414). "As we use the term relevancy, *a certain item* is said to be *relevant to the theme* (which may well be a plan of action or a pursuit) and also to other items because of their relevancy to the theme" (Gurwitsch, 1964, p. 342). This means that the content in the thematic context is not the center of interest, but is the ground of that center. It also means that the content in the thematic context hangs together in a unified way, such that further elucidation of the thematic context would reveal additional relevant content. This context can be attended to as compact, diffuse, clear, enlarged, novel, etc. However, content in the thematic context is not presented as thematic. This would take a transformation of attention, as when step three of a mathematics problem becomes thematic when step two is completed, but was attended to contextually previously.

In contrast, marginal attending is directed toward co-present content that are not materially relevant to the theme and thematic context.[2] "We restrict the term margin to denote such data only which, although co-present with, are irrelevant

---

[2] It might seem that marginal attention is *merely* accessory to thematic attention, and so is dispensable. But the margin is always presented, there is always marginal attention. In the margin, we attend peripherally to the stream of consciousness (or flow of attention), embodied existence, and the perceptual world in general, and these three ever-present orders of existence form the ultimate experiential context for human being. A certain segment or sector of each of these three orders of existence are always attended to marginally, if not thematically. Gurwitsch (1985, p. 51) writes, "In all cases, the presence before consciousness marginally of the three mentioned orders of existence consists in some pertinent data being actually and originally, although not thematically, experienced with their pointing references to a wider context. As long as it is a question of marginal consciousness, this is all that is given."

Making explicit what I see implied in Gurwitsch's (1985) work, I would add a note on how marginal attention works. Attention to marginal content also has a central/contextual structure. A certain segment or sector, called the *halo*, of each order of existence is always marginally attended to within the context of the order of existence as a whole, called the *horizon* (Gurwitsch, 1966, p. 268). For example, while pruning flowers in the garden when I marginally attend to the hedge and sky, these halo presentations have pointing references to the world in general, however diffuse or indiscriminate that larger context or horizon may be. The same holds for the marginal attention to the current moment as part of a flow of attention and the kinesthetic sense as pointing to the fact of my corporeity. The halo is that part of the margin that most closely adjoins the thematic context, and might be relevant to the theme under another perspective (Embree, 1985, xxx).

to, the theme. *Margin*...denotes a *domain of irrelevancy*" (Gurwitsch, 1964, p. 344). The itchy nose in the beginning of the example is merely co-present with but irrelevant to the theme of the coffee spill on the desk. Marginal attending also allows the presentation of Gestalts as consolidated content; and there is a segregation of this content from the theme and thematic context, and from the rest of marginal content. "If marginal consciousness has an additive structure, the reason is that the potential themes, each one of them a well articulated whole, having no relevancy to another, are not interlinked into a unity encompassing all of them by virtue of their respective contents" (Gurwitsch, 1966, p. 270). This segregation is denoted by "irrelevance" (Gurwitsch, 1966, p. 270; 1964, p. 414; 1985, p. 40). But this term stresses especially the absence of a material relation of content between the theme and margin (and so also between the thematic context and margin), and must not be understood to mean that the margin is of no importance.[3]

II. Theme and Margin: Experimental Evidence

What I have looked for in the cognitive science literature are reports that show the kind of thematic context or marginal attentional activity described above. The dominant research paradigms restrict the use of the word "attention" to processing of the focus, target, or theme. So I looked for evidence of processing that is outside of the focus of attention and either relevant or irrelevant to this focal processing; respectively, this would denote contextual or marginal attending. In short, if I am correct in the differentiation of the three types of organizational principles for the three dimensions of attention, this non-focal processing would have to fit the description of either contextual or marginal attending given above.

There are various ways that cognitive scientists may refer to content that is processed outside of the focus of attention but nonetheless has an effect on the processing in the focus of attention. For example, an "unattended stimulus" is not the target that the subject is supposed to attend to, but is somehow "automatically" processed. This non-targeted content can affect subsequent targets such as in the splitting of auditory attention, sometimes called "a Broadbent filtering effect" (see Broadbent, 1958) or it can affect the co-presented target (e.g., Rauschenberger and Yantis, 2001, p. 1259; Eastwood, Smilek, and Merikle, 2001; Posner and Peterson, 1990, p. 29, 31). I will suggest that "unattended" content is better described as content that is attended to

[3] Gurwitsch also uses the terms "indifference" (1966, p. 271; 1964. p. 344) and "disconnectedness" (1985, p. xliv, p. 50), and calls the margin the "domain of contingency" (1964, p. 414).

contextually (i.e., thematic context attending) or attended to marginally, depending on the relevance or irrelevance for the target.

Finding evidence for marginal attention in experimental research is relatively easy, even though it is not called that. Research paradigms assume and preserve William James's distinction between that which is presented as focal and that which is presented as marginal in attention (James, 1983, pp. 19-22; Arvidson, 1992b). Countless experiments use the typical methodology of distractors presented outside the target area and then measure the degree of interference with focal attention caused by the distractors. The distractors are thought to be "unattended" since they are not thematic. A more positive and direct description is that these distractors are attended to marginally, that is, as irrelevant (cf. Gurwitsch, 1964, pp. 282-3; 1966, p. 272).

An invalid cue is also marginally attended at the moment it transforms from valid to invalid. For example, in what became a classic methodology for visual spatial attention research, Posner (1980) found that the appearance of an object on a screen prior to the appearance of the target in the same place on the screen cued or sped up reaction time to the target. This is called a valid cue or prime and is discussed below since it provides context for the target. But a cue that appeared in a different spot than the subsequent target, an invalid cue, slowed reaction time to the target. This cue is invalid precisely at the moment that the target appears somewhere else. Attention to the cue has transformed from being relevant to the appearance of the target (expectancy) to irrelevant to it (merely co-present), and is attended to as irrelevant.

The point here is two-fold. First, experimental psychologists of attention create volumes of research on the assumption that there is a margin in what is presented for the subject. Second, they call the content in this dimension "unattended stimuli" or irrelevant information, and the dimension itself is where "task-irrelevant processing" occurs. A more descriptive and global view of attention would enlarge the description of this dimension and this processing beyond a simple negation of focal attention, and refer to it as marginal attention. Again, this means that this distracting or neutral information is processed outside the focus of attention and is not relevant to that focus; the subject attends to it as irrelevant.

One could object that the word "attention" is inappropriate here, since it connotes a more or less clear and central target and the margin is relatively diffuse and peripheral. For what kind of attention is so-called marginal attention if it is defined precisely as *in*attention to the theme? This objection has some merit, and my response will not sway all readers. Marginal content is peripheral and irrelevant only in relation to the given theme. So the first point is that if the margin is the domain of *in*attention, this negation of attention is already a related process that occurs along with thematic attention. Is it better to call this simultaneous marginal processing memory? Imagination? Perception? Cognition? Unconsciousness? Consciousness? None of these has more to recommend it than a precise distinction between these two types of attention, focal and marginal. The virtue of calling it marginal attention is that this phrase preserves and expresses the fact that this processing is intimately related with thematic attending (as presenting content irrelevant to the object of the current thematic attending) and occurs simultaneously with it under the general heading of attention.

The second point is that this marginal dimension is assumed to exist in attention experiments, and is critical to existing methodology, and is even implicitly codified (as task-irrelevant processing). It has something to do with attention, but is not conventional attending. For example, in a measure of reaction time to the target, subjects in a wide range of experiments are typically instructed to ignore a potential distractor on a video screen. Essentially, they are being asked to attend to it as irrelevant. Of course, the way this is now interpreted is that the subjects did not pay attention to the distractor at all, although it may have affected attention in some way, for example, taxed focal attentional capacity (Arvidson, 2003a). But in measures of vigilance and interference effects, the point is often that the subjects have constantly to make this content irrelevant to the target. This tension between the theme and margin most clearly reveals two distinct dimensions of attention. The subjects are being asked to attend focally to the target and to attend marginally to the distractors.[4] So using the term "marginal attention" is preferable because this type of processing is intimately conjoined with thematic attention, it is more reasonable than other faculties terminology such as memory or unconsciousness, and it is at least implicitly codified as an essential part of the processing that occurs and is measured in many standard cognitive science experiments on attention.

III. Thematic Context: Experimental Evidence

Many metaphors used in attention research preclude meaningful discussion of what I have called contextual attention. For example, the still popular "spotlight" metaphor of attention has a long history in philosophy and psychology, which includes Edmund Husserl (1982, §92), as well as Posner (1980) and Treisman and Gelade (1980), and more recently Baars (1997) and others (e.g., Ohman, Flykt, and Esteves 2001). Gurwitsch (1966, pp. 202, 205) critiqued it in 1929—fatally, in my opinion—but see also LaBerge (1995) and

---

[4] The measure of success or failure on the part of the subject to keep the experimenter directed content distinguished in each of these dimensions of attention (in the way appropriate to each) is called "reaction time," and more specifically, an "interference effect."

Arvidson (1996). The main problem is that a spotlight metaphor can reveal only two dimensions in attention, namely, a focal attention corresponding to what is illuminated, and a non-focal attention corresponding to what is non-illuminated. Not only is the possibility of context as a distinct dimension of presented organization in attention completely precluded by this metaphor, but it is impossible to account for the general richness of attention processes, such as transformations in attention. Gurwitsch believes that attention has more to do with the content of what is attended to, that is, its organization in any given case and its transformation of organization, than with the one dimensional on/off, illuminated/unilluminated metaphor (Gurwitsch, 1966, p. 222; Arvidson, 2000).[5]

One of the most exciting advances in attention research is the attempt to measure the temporal aspect of attention. It is exciting for two reasons. First, it assumes that attention is a dynamic process, and this is exactly the heart of the definition of attention that I argue for, and that Gurwitsch first identified in 1929 with his "General Transformation Law" (1966, p. 248). Second, it forces the researcher to consider the effect of non-focal, contextual information on the target.

For example, Peterson and Kramer (2001) try to define the mechanisms by which contextual cuing guides attention. The ease with which one can use the term "contextual cuing" in a mainstream cognitive science journal article appears remarkable given the paucity of acknowledgment of context as operative in the presentation of a target (rather than say, "unattended stimulus" or "irrelevant information"). In defining the mechanisms for contextual cuing, one of the components they operationally define is recognition. "Recognition entails the identification of the *current context* of the scene and the matching of this information with previously stored instances of the familiar context." (1239-40, emphasis added). These authors explicitly point out that the subject recognizes the scene contextually or, as I would put it, attends to it contextually.

---

[5] Perhaps realizing this, some researchers (e.g., Mangun & Hillyard, 1988) tried to account for context but keep the spotlight metaphor. By postulating a gradient of attention, the co-presented stimuli which constitute the context for the theme or focus are illuminated less brightly than the focus. This sort of gradient in the field of attention corresponds well also with a zoom-lens model of attention. Yet the dimensional difference between thematic attention and contextual attention is not demarcated by a gradient. There is certainly a gradient of attention *within* each dimension. For example, there is this sort of relation of emphasis between formative and formed constituents of the theme, and between contents in the thematic context that are nearer to the theme and those that are more remote, and the same with the margin. But these are intra-dimensional, and a gradient is not sufficient to account for inter-dimensional differences, like that between theme and thematic context (Arvidson, 1992b).

They found that the recognition and achievement of the target were intimately related: "That is, when recognition occurred, the eyes were guided directly to the target; otherwise, the eyes went to another item in the display, with no bias toward the target" (p. 1249). The target is achieved (in thematic attention) only along with contextual attention—attending thematically to some content is always accompanied by attending contextually to relevant content which forms its background as a thematic context.

In "Temporal Contextual Cuing of Visual Attention," Olson and Chun (2001) report that temporally attending to the context can facilitate or cue focal attention. They call the act of paying attention contextually learning. They explain (p. 1300) that "this paradigm, termed temporal contextual cuing, involves searching for a target in a rapidly changing stream of distractors. The sequential organization of the distractors preceding the target contains invariant temporal information." Subjects were shown objects on a screen in an invariant temporal sequencing—a rhythm. And although the recognition of the rhythm was implicit, it formed a temporal context for the direction of attention to a particular point in time or a particular location for target achievement. In other words, subjects paid attention to the context, but not in the way they paid attention to targets, and not in the way they paid attention to distractors. The thematic context in which targets appeared was attended to as *relevant* to the target, which is exactly what is meant by "contextual attending." Moreover, these researchers discovered that the part of the temporal sequence that was closer to the target was more salient and provided greater cuing than parts at the beginning of the sequence, although it does not provide all the cuing (p. 1309). This accords with a dominant feature of thematic context presentations noted by Gurwitsch. The content materially closest to the theme is more intensely related to it than that which is farther away or is only implied by pointing references (Gurwitsch, 1964, pp. 338, 353, 379; 1966, p. 205).

In research on attentional expectancy, Barnes and Jones (2000, p. 267) state that they are examining the influence of "serial context" on time judgments. Effectively, this means that the target is given within a relevant context, and that the context is also attended to but in a way different than attention to the target. According to Barnes and Jones (2000, p. 261), "[These dynamic attending models] place greater emphasis on stimulus time relationships as determinants of real time attending and expectancies. Thus, moment-to-moment attending to events such as speech and music is controlled, in part, by their relational properties, e.g., rate and rhythm." The investigation of the effect of "relational properties" on the achievement of the target is the investigation of "attention" to relevant content presented along with the target, which is nothing less than an investigation of contextual attending.[6]

Researchers in the various visual-spatial experimental paradigms are also implicitly or explicitly reporting on processing of relevant content outside of the focus or theme. For example, Eastwood, Smilek, and Merikle (2001) conducted experiments to determine whether the emotion of a face can guide focal attention, even though that emotion is not presented *in* focal attention. They found that information about the emotion of a face can occur outside of the focus of attention and guide the achievement of the target. The subject attended contextually to the emotion of the face, so that it was relevant for the targeted face. The authors (p. 1004) state that "the results suggest that the emotional expression in a face can be perceived outside the focus of attention and can guide focal attention to the location of the face." Of course, following protocol, the authors refer to this "outside the focus" information, which is nonetheless processed, as "unattended" (p. 1012). The title of the *Perception and Psychophysics* article is telling—"Differential Attentional Guidance By Unattended Faces Expressing Positive And Negative Emotion."

In a variety of ways, researchers are examining the effects of global scene or scene context on focal attention. Usually, these experiments are looking for facilitation effects or inhibition effects. Facilitation effects, such as semantic priming, show that the context speeds up achievement of the target, while inhibition effects show the opposite. For example, Mathis (2002) examined the effects of scene context on semantic processing of the target and found that whether the objects fit with the background affected word-categorization performance. Such a "fit" can only be determined through attention to the background *as* background.[7]

In introducing their experiment on emotion and attention, Ohman, Flykt, and Esteves (2001, p. 466) state that "to detect threatening events outside the spotlight of focused, conscious attention, there must be perceptual processes that automatically scan and analyze the perceptual field." In reporting their

---

[6] Barnes & Jones (2000) distinguish two kinds of "local context": experimental or probabilistic cueing, as in the classic Posner (1980) experiment where subjects expected a target to appear at particular spatial location because that location was cued as most probable and pattern-based relationships that are attended to as deterministic. Of these latter, Barnes & Jones (2000, p. 257) state that "pattern-based approaches to expectancy go beyond local cues and first-order conditional probabilities to view contextual information in terms of relationships among features or elements."

[7] In introducing the article the author sums controversies surrounding the effects of scene context on object identification. The controversies themselves cannot be delved into here. But the very fact that there are controversies of context effects, and that experimenters are working on processing that I would call contextual attending is significant.

experiment, they use the term "surroundings" to refer to what is presented outside the focus, but still relevant. But their preferred term for what I would call contextual attention is "attention control settings." Other researchers (e.g., Folk, C. L., Remington, R.W., Johnston, J. C. 1992) also use this phrase to refer to preparatory or coincident emotional or attitudinal orientations that affect focal attention. The "setting" or setting-up for what is attended to thematically is the thematic context. In addition to fear-relevant stimuli with which Ohman *et al.* were concerned, they observe that (2001, p. 475) "one would expect other emotional or motivational states to involve different attention control settings, for example, resulting in a bias for food-relevant items in the surroundings when one is hungry." The point here is that attending to the context as relevant for whatever might present itself as edible—an attention control setting—is contextual attending.

Smith, Bentin, and Spalek (2001, p. 1289) define semantic priming as "the facilitated processing of a target word when it is preceded by a related prime word." Notice the assumption of relation, so that there is a contextual or categorial relevance between the two contents. Since this serial shifting transformation of attention occurs all within the same thematic context, the transformation is facilitated (e.g., Gurwitsch, 1966, pp. 231-232; Barnes and Jones, 2000; Olson and Chun, 2001; Chelazzi, 1999). This popular experimental paradigm often refers to this contextual attending as "activation of the semantic system." It is now well established within this methodology that semantic priming is reduced or eliminated if the prime task directs attention to the letter level (Smith, Bentin, and Spalek, 2001, p. 1289-90). In other words, when the context is shifted and hence is made irrelevant (now a letter instead of a word), this singling out reduces or eliminates semantic priming because there is now a reduced or eliminated relevance of the semantic context for the letter. Contextual attention has not disappeared, but a new relevance (a new context) is co-presented with the new theme (Arvidson, 1992a). The previous thematic relevance of semantics is now irrelevant, and so semantic priming is diminished. The point is that this robust experimental area implicitly assumes contextual attending.

In order to investigate the process and control of attention, a common experimental paradigm examines interference effects that use distractors that are relevant to a target. One of the documented effects of interference from relevant information that is presented outside of the focus of attention is called the negative priming effect. In negative priming, the theme as target will be achieved more slowly if the current target is relevant for a previous target that was to be ignored. For instance, the word "BUS" is presented at t1, and the word "NURSE" is presented as a distractor (that is, it is attended to marginally). At t2, the achievement of the word "DOCTOR" as target is slowed. The relevance between the words "NURSE" and "DOCTOR" interfere with achievement of the target at t2 because that category of words (or objects) was actively marginalized at t1. I give a more thorough account of negative priming effects elsewhere (Arvidson, 2003a). The salient point here is that this experimental paradigm and the very possibility of an interference effect here rely on the *relevance* of the current target to that which is not attended to focally but is nonetheless attended to, namely as contextual.[8]

Kane, Bleckley, Conway, and Engle (2001) investigate the relationship between the active maintenance of information outside the focus of attention, which experimental psychologists usually call "working memory" and the control of attention. They argue that working memory capacity is a function of "controlled attention" (see also, e.g., Rockstroh and Schweizer, 2001, p. 31; and McElree, 2001). This is a larger view of attention than usually discussed in current research since "controlled attention" is responsible not only for the achievement of the target but also for determining what is relevant or irrelevant to the target—what I have called contextual and marginal attending. "By 'controlled attention' we generally mean an executive control capability; that is, an ability to effectively maintain stimulus, goal, or *context* information in an active, easily accessible state in the face of interference, to effectively inhibit goal irrelevant stimuli or responses, or both" (Kane, et al., 2001, p. 180, emphasis added). This is a clear appeal to attention outside the focus of attention, whether it is called "controlled attention" or "working memory" matters little. The distinction is being made, within the experimental literature, between the theme, and that which is attended to as relevant to the theme (the thematic context), and that which is attended to as irrelevant to the theme (the margin). Implicitly referring to all three dimensions, Kane, et al (2001, p. 170) conclude "Thus, coherent and goal oriented behavior [thematic attention] in interference-rich conditions [marginal attention] requires both the active maintenance of relevant information [contextual attention] and the blocking of or inhibition of irrelevant information [marginal attention]."

In concert with other researchers working on expectancy and the temporal aspect of attention, McElree (2001, pp. 819; 831) identifies the non-attended stimulus as being represented in working memory. McElree (2001, p. 817) writes, "whenever information exceeds the span of attention, successful execution of a cognitive operation requires shunting information between

[8] The title of a recent article investigating this type of interference includes the notion of related or relevant items that are not the theme—Dean, Bub, & Masson (2001) "Interference from Related Items in Object Identification." The authors, however, refer to "dynamic, memory-based representations" instead of acknowledging what I would call contextual attention.

memory and focal attention." So there is a dichotomy between focal attention and working memory, such that whatever occurs outside the focus but affects "executive control processes" occurs in working memory. I think that it is better to say that whatever occurs outside the theme but is relevant to it occurs in the thematic context. This expression is better because it does not invoke some other faculty secondary to attention (like memory) and most importantly because it better describes what Peterson and Kramer (2001) call "recognition," which is a co-processing of the current context of the theme, along with the theme. Experiments show that what is in working memory affects the target or task (e.g., Dean, Bub, and Masson, 2001; Marks and Dulaney, 2001). To the extent that this content gives light, perspective, or orientation to what is currently achieved in thematic attention, this content is part of the thematic context (Gurwitsch, 1964, p. 359) and may facilitate or prime thematic attention. Serial-shifting tasks are a good example of this (Chelazzi, 1999).

Neurological studies are beginning to blur the conventional distinctions between what is attended and unattended, sometimes with tellingly odd expressions. For example, Downing, Liu, and Kanwisher (2001) found that there are attentional effects seen in cortical areas related to task irrelevant stimuli (p. 1336). Given the one-dimensional model of target vs. non-target stimuli, where there is no room for the concept of context (Gurwitsch, 1964, p. 17), the processing of attention in a second dimension is described as "...the resulting spread of attention to irrelevant stimuli" (p. 1336). This expression is curious since attention has "spread" beyond the focus or target. Note that the stimuli are "irrelevant" because they are off-task, according to the view of the experimenter. In fact, the experimenter in this paradigm cannot know if the stimulus that is attended to outside of the focus is relevant or irrelevant from the perspective of the subject since no sense of attentional context is allowed in the paradigm and the subject is not queried. Still, this type of neuropsychological experiment shows that there is attention to something outside of the focus of attention.

In a review of recent neurological studies of attention, Driver, and Frackowiak (2001) are more to the point concerning the presence of relevant non-focal attentional processing. They (2001, p. 1259) write, "moreover, rather than the classically proposed sequence of initial pre-attentive stimulus processing, followed by attentional modulation at a later level, it is now clear that attentional modulation at multiple levels can start *before* (in time) any stimulus processing arises. Attentional influences can thus actually precede what was traditionally considered to be 'pre-attentive' sensory processing!" One does not have to completely share these authors' exuberance about the attentional timeline to recognize that they are *at least* claiming that neurological evidence shows that information *related* to the target (attentional influences), but which is "unattended," is processed along with the target.

Conclusion

By allowing phenomenology and cognitive science to turn toward each other in a friendly way, this paper has tried to articulate a greater meaning and role for the phenomenon of attention in both of these approaches to human nature. I have argued that cognitive science supports Gurwitsch's central insight. This support is implicit and sometimes explicit. The framework was possible because Gurwitsch's "consciousness" has been interpreted as "attention," and cognitive science's "processing outside of the focus of attention" was interpreted as "attention." It is not hard to imagine that Gurwitsch (and other phenomenologists) and cognitive scientists might bristle at these interpretations. Nonetheless, for reasons outlined above, such a move yields mutual illumination between these two kinds of approaches. I think that responsible but creative flexibility in defining attention is needed to get at a phenomenon more primordial and revealing of human nature than "consciousness" as it is now defined.

References

Arvidson, P. Sven. 1992a. "On the Origins of Organization in Consciousness," *Journal of the British Society for Phenomenology* 23: pp. 53-65.

—. 1992b. "The Field of Consciousness: James and Gurwitsch," *Transactions of the C. S. Peirce Society* 28: pp. 833-856.

—. 1996. "Toward a Phenomenology of Attention," *Human Studies* 19: pp. 71-84.

—. 1997. "Looking Intuit: A Phenomenological Analysis of Intuition and Attention, " in *Intuition: The Inside Story: Interdisciplinary Perspectives,* eds. Robbie Davis-Floyd and P. Sven Arvidson. New York: Routledge.

—. 1998. "Bringing Context into Focus: Parallels in the Psychology of Attention and the Philosophy of Science," *Journal of Phenomenological Psychology*, 29: pp. 50-91.

—. 2000. "Transformations in Consciousness: Continuity, the Self and Marginal Consciousness," *Journal of Consciousness Studies* 7: 3-26.

—. 2003a. "A Lexicon of Attention: from Cognitive Science to Phenomenology," *Phenomenology and the Cognitive Sciences,* 2: forthcoming.

—. 2003b. "Moral Attention in Encountering You: Gurwitsch and Buber," *Husserl Studies* 19: 71-91.

Baars, Bernard. 1997. *In the Theater of Consciousness: The Workspace of the Mind.* Oxford: Oxford University Press.

Barnes, Ralph and Mari Riess Jones. 2000. "Expectancy, Attention, and Time," *Cognitive Psychology* 41: 254-311.

Broadbent, Donald E. 1958. *Perception and Communication.* London: Pergamon.

Chelazzi, Leonardo. 1999. "Serial Attention Mechanisms in Visual Search: a Critical Look at the Evidence," *Psychological Research*, 62: pp. 195-219.

Dean, Michael, Daniel Bub, and Michael Masson. 2001. "Interference from Related Items in Object Identification," *Journal of Experimental Psychology: Learning, Memory, and Cognition*, 27: pp.733-743.

Downing, Paul, Jia Liu, and Nancy Kanwisher. 2001. "Testing Cognitive Models of Visual Attention with fMRI and MEG," *Neuropsychologia*, 39: pp. 1329-1342.

Driver, Jon, and Richard Frackowiak. 2001. "Neurobiological Measures of Human Selective Attention," *Neuropsychologia*, 39: pp. 1257-1262.

Embree, Lester. 1985. "Editor's Introduction," Aron Gurwitsch, *Marginal Consciousness* (pp. xi-xlii). Lester Embree (ed). Athens, Ohio: Ohio University Press.

Eastwood, John, Daniel Smilek, and Philip Merikle. 2001. "Differential Attentional Guidance by Unattended Faces Expressing Positive and Negative Emotion," *Perception and Psychophysics*, 63: pp. 1004-1013.

Folk, Charles, Roger Remington, and J.C Johnston. 1992. "Involuntary Covert Orienting is Contingent on Attentional Control Settings," *Journal of Experimental Psychology: Human Perception and Performance*, 18: pp. 1030-1044.

Gurwitsch, Aron. 1964. *The Field of Consciousness* (Pittsburgh: Duquesne University Press).

—. 1966. *Studies in Phenomenology and Psychology*. Evanston, IL: Northwestern University Press.

—. 1985. *Marginal Consciousness*. Lester Embree (ed.). Athens, Ohio: Ohio University Press.

Heidegger, Martin 1962. *Being and Time*. John Macquarrie and Edward Robinson (trans.). New York: Harper and Row.

Husserl, Edmund. 1982. *Ideas Pertaining to a Pure Phenomenology and to a Phenomenological Philosophy: First Book*. Fred Kersten (trans.). The Hague: Nijhoff.

—. 1991. *On the Phenomenology of the Consciousness of Internal Time (1893-1917)*. John Barnett Brough (trans.). Dordrecht: Kluwer.

James, William. 1983. *Talks to Teachers on Psychology and to Students on Some of Life's Ideals*. Cambridge, MA: Harvard University Press.

Kane, Michael, M. Kathryn Bleckley, Andrew Conway, and Randall Engle. 2001. "A Controlled-Attention View of Working-Memory Capacity," *Journal of Experimental Psychology: General*, 130: pp. 169-183.

LaBerge, David. 1995. *Attentional Processing: The Brain's Art of Mindfulness*. Cambridge, MA: Harvard University Press.

Mangun, George and Steven Hillyard. "Spatial Gradients of Visual Attention: Behavioral and Electro-physical Evidence," *Electro-Encephalography and Clinical Neurophysiology*, 70: pp. 417-428.

Marks, William and Cynthia Dulaney. 2001. "Encoding Processes and Attentional Inhibition in Directed Forgetting," *Journal of Experimental Psychology: Learning, Memory, and Cognition*, 27: pp. 1464-1473.

Mathis, Katherine. 2002. "Semantic Interference from Objects Both in and out of a Scene Context," *Journal of Experimental Psychology: Learning, Memory, and Cognition*, 28: pp. 171-182.

McElree, Brian. 2001. "Working Memory and Focal Attention," *Journal of Experimental Psychology: Learning, Memory, and Cognition*, 27: pp. 817-835.

Ohman, Arne, Anders Flykt, and Francisco Esteves. 2001. "Emotion Drives Attention: Detecting the Snake in the Grass," *Journal of Experimental Psychology: General*, 130: pp. 466-478.

Olson, Ingrid and Marvin Chun. 2001. "Temporal Contextual Cuing of Visual Attention," *Journal of Experimental Psychology: Learning, Memory, and Cognition*, 27: pp. 1299-1313.

Peterson, Mathew and Arthur Kramer. 2001. "Attentional Guidance of the Eyes by Contextual Information and Abrupt Onsets," *Perception and Psychophysics*, 63: pp. 1239-1249.

Posner, Michael. 1980. "Orienting of Attention," *Quarterly Journal of Experimental Psychology*, 32: pp. 3-25.

Posner, Michael and Steven Peterson. 1990. "The Attentional System of the Human Brain," *Annual Review of Neuroscience*, 13: pp. 25-42.

Rauschenberger, Robert and Steven Yantis. 2001. "Attentional Capture by Globally Defined Objects," *Perception and Psychophysics*, 63: pp. 1250-1261.

Rockstroh, Sybille and Karl Schweizer. 2001. "Contributions of Memory and Attention Processes to Cognitive Abilities," *The Journal of General Psychology*, 128: pp. 30-42.

Smith, Marilyn, Shlomo Benton, and Thomas Spalek,. 2001. "Attention Constraints of Semantic Activation During Visual Word Recognition," *Journal of Experimental Psychology: Learning, Memory, and Cognition*, 27: pp. 1289-1298.

Treisman, Ann and G. Gelade. "A Feature Integration Theory of Attention," *Cognitive Psychology*, 12: pp. 97-136.

Chapter 9

The Structure of Context and Context Awareness

David Woodruff Smith
University of California, Irvine

Abstract

The role of context in cognition or intentionality has been studied in disciplines ranging from phenomenology to philosophy of mind and language to cognitive science. Here I propose to appropriate and adapt parts of Aron Gurwitsch's analysis of "marginal" and "thematic" consciousness. Gurwitsch factored the "field of consciousness" into Theme, Thematic Field, and Margin, and he distinguished three "orders of existence" in Stream of Consciousness, Embodied Existence, and Perceptual World. I propose to rebuild these distinctions as linked structures of intentionality: formal meaning (noematic) structures defining focus of attention, and formal objective structures defining context of experience. The analysis will clarify results in cognitive science concerning "inattention blindness." My larger aim is how to analyze intentionality as a "formal" ontological structure within a framework of formal ontology.

I. Contextual Consciousness

As Edmund Husserl launched phenomenology in his *Logical Investigations* (1900-01) and *Ideas I* (1913), a crucial part of his analysis of intentionality was an account of the "horizon" of the object of experience. Husserl's results were extended and modified by Aron Gurwitsch in *The Field of Consciousness* (1964) and *Marginal Consciousness* (1985). Gurwitsch's analysis of "marginal" consciousness sharply refined elements in Husserl's analysis of horizon, and his results prompt the present study.

In "marginal" consciousness we are aware of things in the larger context of experience, things outside the focus of attention. Our present concern is the *structure* of the context of experience and our awareness of the context of our experience.

As we proceed, we shall find ourselves leapfrogging over Gurwitsch as Gurwitsch leapfrogged over Husserl.[1]

---

[1] In the interests of disclosure: I propose to appropriate and reconstruct key ideas from

L. Embree (ed.), Gurwitsch's Relevancy for Cognitive Science, 169-186.

II. The Role of Context in Intentionality

A large bird circles overhead, shifting its balance on the wind currents. My perception *cum* judgment of the bird aloft has a phenomenological structure that we may begin to articulate in the following phenomenological description:

In this experience I see that bird and judge that it is a turkey vulture. The *indexical* content or sense "that," or "that bird," in my visual experience intentionally prescribes the object actually now there visually before me and affecting my eyes, that is, in the present *context* of experience (See Smith 1989 for details of this analysis).

An instructive model of indexical intentionality is that of linguistic indexical reference. When I say, "That is a turkey vulture," my speech act, using the *indexical* term "that," depends on my perceptual act of seeing what I am referring to, and forming a judgment about it. Thus, the object of reference is an object in the context of my perception. Indeed, Husserl and Gurwitsch both addressed the relation between the "occasional expression," "this" or "that," and visual awareness of the object referred to on the given occasion. Our concern here, however, lies with the form of indexical awareness rather than indexical language.[2]

The structure of *indexical awareness* in perception—of intentionality via *indexical sense* invoking context—I have appraised elsewhere (Smith 1989), drawing on the phenomenology of perception and the semantics of indexical expressions. Here, drawing on Gurwitsch's results, I want to develop a closer account of the structure of context and contextual awareness in vision, articulating the *form* of perceptual consciousness of environment, specifying just how the content of perception calls on the context of perception. The model we will

---

Gurwitsch and to put them to work within a particular framework for synthesizing phenomenology, ontology, and cognitive science. This approach will call for modifying some of the broad "transcendental" themes dear to the hearts of Husserl and Gurwitsch in their era. My basic intepretation of Husserl on intentionality, noema, and horizon is that in Smith and McIntyre 1982. My analysis of indexical awareness and the role of context and indexical content therein is that in Smith 1989. My take on formal ontology and its role in phenomenology is that of Smith 1995. My approach to naturalism in contemporary philosophy of mind *cum* cognitive science (we need a wider conception of "nature" that accommodates consciousness, intentionality, meaning, and culture) is that of Smith 1999. Relevant background discussions are found in two collections that address phenomenology and cognitive science (at two times in the evolution of "cognitive science"): Dreyfus (ed.) 1982 and Petitot *et al.* (eds.) 1999.

[2] During the 1970's David Kaplan developed a well-known model of the logic of demonstratives and other indexical expressions, published in his "Demonstratives" (1989). Husserl had sketched a similar model already in *Logical Investigations*, I, §26, VI, §5. Gurwitsch addressed Husserl's model in an appendix to *Marginal Consciousness*. Kaplan's and Husserl's accounts of demonstratives, and the related issues of intentionality, are studied in Smith 1982.

develop helps to address issues of how consciousness and context interact in vision, including the problem of "inattention blindness" in recent cognitive science. Ultimately we shall address the formal ontology of intentionality and the role of context therein.

III. Gurwitsch's Phenomenology of "Marginal" Consciousness

When I see "that bird" overhead, I am aware of much more than the bird. When I see its tilting flight and take it for a turkey vulture, I am to some degree aware of the sky, the clouds, the treetops, my pointing finger, my watching the bird, my companion's watching the bird, and so on. What is the *structure* of my overall awareness of this ensemble of things?

William James spoke of the "fringe" of consciousness, the range of things which are outside the focus of attention. Edmund Husserl spoke similarly of the "horizon" of consciousness, including (in first-person idiom) what I am only implicitly aware of in an experience: here including the back side of the bird I see, the air through which it flies, the world around it and around me and my companion, the *Umwelt*, and reaching into the possibilities left open in regard to the object as experienced. Gurwitsch subsequently developed a specific analysis of the structure of what he called the "margin" of consciousness. We may summarize that analysis as follows. (We draw details from *Marginal Consciousness* set against the results of *The Field of Consciousness*, hereafter *MC* and *FC* respectively, guided also by Lester Embree's overview of Gurwitsch's system in the Editor's Introduction to *MC*.)

In every act of consciousness, Gurwitsch held, the subject is presented with a "field of consciousness." What psychologists call the "visual field" is only a special case. Now, this field has a specific structure. The *field of consciousness* divides into three parts: the theme, the thematic field, and the margin. The *theme* is the object (individual, event, or whatever) that is the focus of attention (in vision or thought or volition or whatever). The *thematic field* is the array of things that are co-present (also "meant") in the field and relevant to the theme. The *margin* is the array of things that are co-present in the field but not relevant to the theme. (*FC* 4, 55-56, *MC* xvii ff.) Broadly, the "margin" would be what the act presents as surrounding the object of attention. Gurwitsch first addressed items of awareness that are relevant to the object and then turned to items not relevant, calling the latter "margin" proper, detailing (in *MC*) "marginal" awareness of one's stream of consciousness, one's body, and one's perceptible spatio-temporal environment.

Take the case of my seeing "that bird" overhead. The *field* of my perception is organized around the bird itself, the focus of my attention. That object is the *theme* in my visual field. The *thematic field* (in the overall field) includes a variety of things that are also present in the field and relevant to the bird: its wings, beak, featherless head, topside feathers, its being a turkey vulture, its tilting flight; the clouds behind it in the sky; the dark green treetops of Douglas fir; the grass hillsides below; and so on—all elements of the perceived circumstance as meant in this particular experience, but elements that are relevant to the sense of "that" as a bird, a vulture, floating on wind currents, etc. Yet there is still more to the field of my perception. The *margin* in the field includes a variety of things that are also present in the field but are not relevant to the theme: the bicyclists pedaling along the road nearby, the direction of the road heading into town, the grocery store in town, the rising price of milk at the store, etc.—all elements of the present circumstance as meant, as taken in my experience, but elements that are irrelevant to the bird I see, the focus of my attention. In fact, my ever-wider sense of the world includes my background understanding of a whole range of things, from the geography of the continents to the Pope's role in world affairs to economic trends of global capitalism to principles of higher mathematics. But these aspects of my *Umwelt* reach further into the background of my field of consciousness than we shall study here, so let us focus on more obvious features such as in the simple case recounted.

This Gurwitschean analysis is supposed to be perfectly general. Visual perception is only a special case, in which we observe a familiar structure of the visual field—wherein classical phenomenologists (Husserl, Gurwitsch, Merleau-Ponty) detailed far more meaningful content than a mere distribution of colors or even a minimal interpretation of what I see as a bird or even a turkey vulture. Every act of consciousness, on this analysis, presents me with a *field of entities* organized as theme, thematic field, and marginal (non-thematic) field. The guiding assumption is that in every (typical) act of consciousness I am *attending* to something while also *aware* in different ways of other things related to the focus of attention. Some of these peripheral things are semantically ("noematically") relevant to the theme, but others are not. Gurwitsch's distinctions we take thus to define an *invariant structure* in the field of things presented or intended in an act of consciousness.

Gurwitsch's analysis of the field of consciousness was developed within a basically Husserlian theory of intentionality, and we must interpret the parts of the "field" accordingly. We are concerned at this point not with the object of consciousness and its environment *per se*, the things as they are outside my experience. Rather, we are presently concerned with these things *as* I experience them, as they are "intended" or meant in this particular act of perception. In the Husserlian idiom adopted by Gurwitsch, we are concerned with phenomenological structures of experience, structures of the intentional content of my perceptual experience, called the meaning or sense or the *noema* of the experience.

Accordingly, we shall assume the following broadly Husserlian analysis of the structure of intentionality:

background |— I — act — content ——> object — environment .

We approach Gurwitsch's results within this framework.

We should think of the Gurwitschean distinction among three parts of the field of consciousness, I propose, as a formal or "logical" distinction in the noematic meaning structure of an act of consciousness. Thus, the sense of an act is divided into three parts defined by these intentional or semantic *forms*: Theme, Thematic Field, and Margin. Every act's sense is defined by the *form* Field-of-Consciousness, and that form is structured syntactically into these three forms. Instead of looking to the propositional form of "Subject + Predicate," on this view, we look first to the thematic form of "Theme + Thematic Field + Margin." Indeed, a propositional form takes over only after an object is abstracted from such a field: I see "[Theme:] that" and judge that "it is a turkey vulture."

There are points of difference between Gurwitsch and Husserl, and between my ensuing model and theirs. My task, however, is to draw from Gurwitsch (read against a Husserlian background) the details of noematic structure surveyed just above.

IV. Gurwitsch's Phenomenological Ontology of "Orders of Existence"

With this phenomenology of the "field of consciousness" Gurwitsch joined an ontology that distinguishes three "orders of existence." (*FC* 415, *MC* xxxi, 40ff.)

According to that piece of ontology, there are three distinct "orders" of existence that frame our consciousness: Stream of Experience, Embodied Existence, and Perceptual World. I capitalize these terms to suggest that we interpret them as specifying three categorially distinct domains or forms of being. Each act of consciousness belongs to a stream of consciousness, "my" flow of experience. But "I" am embodied, and so my experiences are essentially related to my body, my embodied self, as I perceive through my eyes and act by moving my arms, legs, etc. Moreover, my acts of consciousness along with myself (this embodied subject of consciousness) exist within a wider world in which I am fundamentally related to things through perception, as well as action. Thus, the world in which my consciousness occurs is organized so as to include *inter alia* three domains: that of my stream of consciousness, that of my embodied existence (which includes my embodied consciousness, from perception to thought and emotion to action), and that of the surrounding world which I engage in perception (and thought and action).

This Gurwitschian ontology may be adapted (for present purposes) as an analysis of three types of ontological structure *oriented* to consciousness. In any act of consciousness, according to Gurwitsch, I have some degree of *marginal* awareness of my stream of consciousness, my body, and my perceptual environment. (That is the running theme of *MC*.) So Gurwitsch's analysis of marginal consciousness leads into his ontology of three orders of existence. However, what we have in that ontology (in my view, with its own particular focus) is really an analysis of the ontology of the *context* of an act of consciousness—as opposed to further analysis of the *content* or meaning in the act. Indeed, Gurwitsch writes:

> *[T]he relationship between the field actually perceived and the surrounding horizon ... is a special case of the continuity of context based upon relevancy. ...* This context between the field actually perceived and the wider environment to which the field belongs and of which it presents itself as a part is experienced through and in the form of horizonal awareness. In other words, *horizonal consciousness is awareness of the context* in question. [*MC* 46]

So we distinguish the context from the awareness of context; the former belongs to ontology, the latter to phenomenology. (The structure of "horizon," we should note, leads into what is merely implicitly intended concerning the object of consciousness, beyond what is "actually perceived."[3])

In stressing the *ontology* of context I am modifying Gurwitsch's approach. When he defines "orders of existence," he places them within the "marginal" noematic structure of consciousness, defining marginal aspects of the object-as-intended. I want to distinguish, however, between such *noematic* structures and the *objective* aspects of context they "intend." Accordingly, I propose, there are three categorial structures of the *context* of an act of consciousness: forms that define the subject's stream-of-consciousness, his or her embodied-existence, and his perceptual-world, including (in the case of perception) the object's relation to the perceptual experience and the embodied subject. In my scheme, these are forms of the act's context itself, not of the act's noema (*cum* horizon). So this analysis of three contextual forms is an analysis of ontological, not phenomenological, structures of act-and-object. But the specified structures of the object are defined by its contextual relationships to consciousness. That is, every act of

[3] The relevant analysis of horizon is a main theme of the later chapters in Smith and McIntyre 1982.

consciousness (in the normal case) occurs within a given subject's stream of consciousness, which is tied into that subject's body or embodied existence, which occurs within a wider environment related to the subject's embodied perceptual experience and action, including social interaction. (Compare *MC,* 41ff.)

Gurwitsch's ontology is of a piece with his conception of transcendental idealism, in Husserl's scheme going beyond Kant's. For him, every object in the world is correlated, and arguably equivalent, with a system of noemata: where a noema is defined as the object *as* intended in a certain way. This view is an updated Kantian scheme, substituting the Husserlian notion of noema for the Kantian notion of phenomenon. For our purposes, I want to bracket those ontological issues of transcendental idealism—and methodological issues of "transcendental reduction." We may draw precisely the Gurwitschean distinctions within the field-of-consciousness and within the *Umwelt* or context-of-consciousness, without committing ourselves to any form of idealism or to any specific transendental methodology. Accordingly, we may import them into a realist ontology of indexical intentionality.

Let us grant the basic details in the above analysis of the field of consciousness and the cognate *Umwelt* of experience. What are we to make of this analysis today? How does it fit into a larger theory of consciousness and its place in the world?

V. Formal Structures of Meaning and Object

In *The Field of Consciousness* Gurwitsch writes:

> The theme-thematic-field-structure is a formal invariant of organization having universal significance. The structure is realized at every moment of conscious life, whatever object (in the broadest sense of the term) presents itself through a given act. As every formal invariant assumes a specific form in every concrete variety in which it is embodied, so is relevancy specified in every case in which a theme appears in a thematic field. As actually experienced, relevancy always is qualified and specified in accordance with, and in dependence upon, the material contents of both the theme and the thematic field. [379]

In *Marginal Consciousness* Gurwitsch further explicated the "marginal" structure of the field of consciousness (presenting things co-present but not relevant). Thus, the structure Theme/Thematic Field/Margin is, according to Gurwitsch, a *formal invariant* structure in every act of consciousness. And so is the structure Stream of Consciousness/Body/Perceptual World (*MC* 40-41). I heartily agree, and I want to stress the importance of this notion of *formal* structure—in phenomenology and ontology.

Logical and linguistic theory commonly analyzes the *formal* "syntactic" structures of expressions in a language, distinguishing, for instance, the *logical forms* of Term, Predicate, Sentence, etc. Logicians who like ideal meanings also recognize parallel formal structures in meanings, distinguishing the *meaning forms* of Individual Concept, Predicative Concept, Proposition. Now, Husserl posited formal structures of meaning in the ideal *intentional contents*, or noemata, of acts of consciousness, extending the notion of *form* from logic to phenomenology. Furthermore, Husserl posited *formal* structures in objects in the world outside language and consciousness, distinguishing the *objective forms* of Individual, Property, and State of Affairs (also Unity, Group, Number, etc.)—he thus called states of affairs "syntactical objectivities." I propose to use such an extended notion of formal structure in order to appraise the neo-Gurwitschean analysis sketched above.

Basically, we have modeled two linked schemes of formal structure. First, there are the formal structures of *context-awareness*: these are the attentional forms of focal consciousness, thematic consciousness, and marginal consciousness. Second, there are the formal structures of the *context* of consciousness: these are the forms of the stream of consciousness, the body or embodied existence, and the surrounding perceptual world. The first are formal *phenomenological* structures of intentional contents or noemata; the second are formal *ontological* structures of objects or entities that define (not just any ontological structure but) the context of consciousness. And the correlations between such forms define the intentional "semantics" of indexical awareness, including my awareness of structured parts of my surroundings when I see "that bird" and judge that "that is a turkey vulture."

This emphasis on the formal structure of context is not alien to Gurwitsch. In his discussion of "essentially occasional expressions"—"this," "here", "I," "now," etc.—Gurwitsch explicitly writes (in the Appendix of *Marginal Consciousness*):

> We have seen that normal essentially occasional expressions have their phenomenological origin in the permanent awareness, at least in marginal form, of the perceptual world, embodied existence, and the stream of consciousness in their mutual interrelations. Correspondingly, *the basis in experience which the demonstratives, as considered here* [when referring to something in the context of discussion rather than in the perceptual surroundings], *derive their status lies in the phenomenon of context as such, i.e., in the fact that every theme*

> *appears in a thematic field, emerging from and occupying the center of this field.* ... [78]
>
> The formal structure of the phenomenon of context does not depend upon the specific nature of any particular context but is an invariant of consciousness. This structure consists in the way a thematic field is organized with respect to a center which stands in relations of greater or lesser immediacy to the items of the field. [79]

I completely agree. But I want to separate two formal structures which do not seem clearly distinguished in Gurwitsch's text (perhaps owing to his transcendental idealism). These are: the formal structure of the *context* of consciousness, and the formal structure of *attention* in consciousness and thus of *context-awareness*.

We are now in a position to join our unfolding account of contextual consciousness with an analysis of indexical awareness in perception (along the lines drawn in Smith 1989). In the resulting model of perceptual intentionality, the indexical structure of seeing "that" will be further analyzed by appeal to the structures of context and context-awareness we have been marking out above.

VI. The Formal Structure of Attention: Noematic Analysis

The ideal content of an act of consciousness—what Husserl called the noema—is structured into two parts: the *sense*, which prescribes or presents the object of consciousness in a certain way; and the *thetic character*, which specifies the act type (say, perceptual) and further modifications of the presentation of the object. The first embodies the *mode* of presentation of the object, specifying its kind, properties, relationships; the second embodies the *modality* of presentation, specifying whether the object is intended in perception, imagination, desire, etc.—and more, as we soon see. Within this distinction of noematic structure we now approach Gurwitsch's distinctions in the field of consciousness. The terminology of "mode" and "modality" of presentation is my updated idiom (in Smith 1989) meant to capture the Husserlian distinction of sense and thetic character.

Return now to the case of my seeing "that bird" overhead.

The sense "that bird" or "that turkey vulture" in my visual experience does not merely specify an individual object, that being circling haltingly overhead. Rather, it specifies that object as part of an ensemble of entities: the *field* of consciousness presented in my perception. This talk of "field" is a natural way of talking about the structure of the content or sense of the perception, rather than the appropriate object(s) *per se*. That is, the field is a *semantic* or *intentional* field, a structure of meaning, not the ensemble of objects themselves in the external world. (We could define the field as the ensemble of external objects, tied to the act of consciousness by intentional relations, but that would be a different approach, by passing the noematic structures in favor of the things they "mean.")

Now, the modality of my experience includes not only the form of *vision*, but also a distribution of *attention* over different things in the field of things presented through the sense.[4] Assuming the distinctions drawn above, there are three *modalities of attention* distributed within the modality and structuring the presentation of entities in the field of consciousness. First, something can be presented *thematically*, that is, *focally* or in the focus of attention. Second, something can be presented *relevantly* or *theme-relevantly*, that is, in the penumbra of attention (the "thematic field") with a relevance defined by the sense of the theme. Or, third, something can be presented *marginally*, that is, in the "margin" of attention (the "marginal field") with no particular relevance defined by the sense of the theme.

For example, in my experience of seeing the bird, the vulture itself is presented *thematically*, in the focus of my attention. Its topside feathers and bald scalp—which I do not directly "see" but implicitly take to be there—are presented *relevantly*, but not thematically. And the bicyclist riding past me is presented only *marginally*. It is not easy to say what counts as relevantly presented.[5] This is because relevancy is to be defined in relation to the particular noematic content modified by the modality "thematically." Thus, the content "bird" in my act implicitly specifies an organism with a head and beak; and, as I take "that bird" to be a turkey vulture, the content implicitly specifies a bald scalp. These elements of meaning are relevant. But the proximity of a bicyclist is not relevant to the content "bird" or "turkey vulture" in my perception: unless it is part of my understanding of turkey vultures that they like to prey on unsuspecting cyclists!

We should treat these noematic structures of attention, I propose, as formal structures of meaning, like logical forms: phenomeno-logical forms. Formal structures are structures that are invariant over appropriate "contents." Thus, atomic sentences in English invariably have the form "Subject + Predicate" or "Term + Predicate," and compound sentences are formed by conjunctions as in the form "P and Q," etc. Similarly, noemata invariably have the form of "Modality + Mode" with three modalities of attention distributed over appropriate com-

---

[4] Compare Husserl, *Ideas I*, §§ 37, 92, on "mode of attention" and the mode "focality." See Smith 1970, placing these in the *Gegebenheitsweise* part of an act's noema, thus in what I have since called the modality of presentation.

[5] See Embree 2003 (this volume) on Gurwitsch's three species of relevancy.

ponents of sense. We might structure the phenomenological description of my experience of seeing the bird in something like the following form, which articulates the phenomenological form of the noema of the act:

> I see [thematically (that bird) + relevantly (its top side) + marginally (that bicyclist)] .

We may schematize the phenomenological description, specifying the form of the noema, in the following *noema schema*:

$$\mathbf{I\ A\ [\ T(\ S_1\ ) + R(\ S_2\ ) + (M(\ S_3)\ )\ },$$

where

I is the ego form ("I ..."),
A is the act-species form (the position for "see"),
[T(__) + R(__) + M(__)] is the form of attention,
which divides into thematic, relevant, and marginal modalities,
$\mathbf{S_1}$, $\mathbf{S_2}$, $\mathbf{S_3}$ are components of sense modified by the attentional components.

If the Gurwitschean analysis of attention is basically correct, then (we are proposing) this formal structure applies to any noema of any act of consciousness. That is the point of the formal noematic analysis of the "field of consciousness."

In the case of vision, there are further structures specific to the "visual field." In particular, objects of sight are presented, in accord with the sense in the noema, as having spatiotemporal extension in their surroundings. And some features of objects in the visual field are presented "sensuously" while others are not. Thus, the black-feathered bottom side of the bird and the shape of its wingspan are presented sensuously, but the top side of the bird is not "sensuously" presented, not seen with visual "fullness" or "evidence," with sensory "qualia." This *sensuous* character is another formal feature in the modality of presentation in perception, but that character is not our main concern here.

When we say that I see "that bird," we emphasize the indexical sense "that," which intentionally prescribes the individual object actually now there (overhead) and affecting my eyes on this occasion, that is, in this context. But the sense "that" (with its uniquely indexical force calling on context) *presupposes* and so depends on the formal structure of attention. For the object designated by the sense "that" in a given experience is the object in the focus of attention. In the present analysis, that is, "that" prescribes its object in the manner specified only if it occupies the thematic position in the full noema, the position "**T** (___)"

as specified in the noema schema above. In this way the analysis of attention underwrites the analysis of the indexical sense "that."

VII. The Formal Structure of Context: Objective Analysis

We might assume that the visual field is a direct pictorial mapping of the objects before one, ultimately a mapping of the distribution of colors and shapes directly visible from one's point of view. Then the content of my seeing that bird would be (in today's digital idiom) a distribution of black points (defining "that bird") amid blue and white points ("sky and clouds") and other points of green ("treetops"). This model of vision follows an extreme empiricism like that of classic sense-data models of vision (from Mach to Ayer). However, as soon as we begin to appraise the "meaning" of things in perception, everything looks different.

The sense "turkey vulture" or even "bird" far outruns the shape of black dots in a particular pattern, not least because the sense signifies a biological species (whose nature and identity scientists may still debate). Moreover, the sense "that" carries significance quite different from a pattern of sensory data like a digitalized television image. Consider the focus of attention: this structural feature of vision is not a matter of color distribution; the black of the vulture does not stand out because it is projected with more intensity on my retina (though the distribution of rods and cones makes a big difference). No, the distribution of attention is not a distribution of light; it is a distribution of *meaning*, of noematic structure, as we have considered above.

There is indeed a "disconnect" between the structure of the perceived *environment* and the structure of the *noema* of the perceptual experience. Grant the appropriate form of sense in the noema, say, "that turkey vulture," and grant the form of attentional modality (Theme, Thematic Field, Margin) distributed over appropriate components of sense: as in the noema schema proposed above. Semantically linked with this noematic structure is an ontological structure that frames the relevant *context* of consciousness, the context invoked by the sense "that turkey vulture." Drawing on the Gurwitschean analysis of three "orders of existence," we can frame an ontology of the context of consciousness. However, that structure of context does not mirror the noematic structure of attention.

The *context* of any act of consciousness, we may thus propose, is framed by three categories of ontological structure: The stream of consciousness, the body, the perceptual environment. These categories do not define anything like basic ontological categories, say, extending a Cartesian or Aristotelian metaphysics. Instead, they define structures in the context of consciousness. This framework is a formal ontological structure of the context of any act of consciousness: it is invariant for all acts, whether perception, judgment, imagination, desire, or whatever, and regardless of whether the sense is indexical. However, this structure of context does not look like a projection of the structure of attention. Rather, it defines ontological parameters relevant to the context of any act of consciousness experienced by an embodied subject consciously living in surroundings accessible by sight, touch, hearing, etc. Well, of course: these are the fundamental aspects of the human being, the basic features on which all acts of consciousness depend in different ways, and to which the sense of typical acts of consciousness may appeal semantically, including an indexical sense structured by attention, such as the perceptual content "that."

Again consider the case of my seeing the vulture. My visual experience is itself part of the relevant context of consciousness: light from the object is impinging on my eyes and partially causing my experience. My body is also part of the context of my experience: I look from and with my eyes, turning my head to see, pointing my finger toward the bird in flight. My visual surroundings also are part of the context of my experience: the world is spread out around me, around my body, around my visual experience, with the bird up there in the sky circling over me on the ground, and so forth. We might define the appropriate context of my experience as the structure: <this visual experience, my body, my visible environment>. This structure is defined by the contextual categories: The stream of consciousness, the body, the perceptual environment. Now, the sense "that turkey vulture" in my visual experience has a specific *intentional force* that may be defined as follows (following Smith 1989):

> The content "that turkey vulture" in this visual experience in this visual environment intentionally prescribes a certain object O if and only if O is actually now there before me in this visual environment and partly causing this visual experience.

Not only does the indexical sense "that ..." prescribe an object in the context of the experience, but that context is itself structured by the ontological forms of the stream of consciousness, the body, the perceptual environment.

Turning to the cognate visual judgment, we may define the *truth conditions* of my visual judgment as follows:

> Where I judge "that is a turkey vulture" in a given context: the content "that is a turkey vulture" in my experience in the present context is true if and only if in the present context there is an object O actually now there overhead and affecting my eyes, and O is a turkey vulture, where the content "that" occurs in the focal position in my vision's noema

"[**T** (that) + **R**(___) + **M**(___)]." [6]

Here we map out the formal structure governing the intentional relation between the act and the state of affairs it represents or "intends." The formal structure of attention partly defines the form of meaning "that" in the propositional content "that is a turkey vulture." And the formal structure of context partly defines the form of my cognitive context featuring the given thought in my stream of consciousness, the physical-causal relation between the bird and my eyes (part of my body), and the spatiotemporal-causal *Umwelt* surrounding me and my consciousness. Although these features are formal, they apply to highly contingent events in our world.

We have been focusing on a simple case of perception, where the content is partly indexical and semantically invokes the context of the perception. But if the context of consciousness is invariably framed by the three categories mentioned, how does context affect intentionality in other cases? Suppose I am thinking that "472 + 39 = 511." How on earth does the context—involving my stream of consciousness, my body, my surroundings—affect the intentionality of my thought? What role has any of the three elements of context?

As Gurwitsch rightly observed (*MC* 39-40), even when doing abstract mathematics one has some degree of marginal awareness of one's surroundings, of one's body, and of one's consciousness. Suppose I was doing the arithmetic in my head and then think "472 + 39 = 511." My thinking is conscious and so includes a certain inner awareness of my thinking. (On my analysis in Smith 1989, that form of *inner awareness* is part of the modality of perception—a story we will not go into here except to note the indexical appeal to context.) Further, while thinking the arithmetic result, my eyes are open and I am vaguely aware of light in the room, the desk before me, etc., even though none of the details of my surroundings matter to me at the moment. Moreover, I have some degree of awareness of my body, as I sometimes count "on my fingers" and in this case (while typing on the computer) I imagine the columns of numbers, that is, in an orientation of up/down keyed to my own body's position. Now, all these awarenesses are *marginal*: to whatever degree I am aware of these items in the structure of attention, they are semantically irrelevant to the arithmetical

---

[6] The scheme of truth conditions familiar in logical theory today developed well after Husserl's time. When we look to conditions of truth and reference for indexical or "occasional" expressions, the notion of meaning must be ramified. In Husserl's and Gurwitsch's account of "occasional" expressions, there are two levels of meaning for, say, "that": the "indicated" meanings in speakers's concrete perceptual acts are governed by the "indicating" meaning, or general meaning function, of the word "that." See the Appendix of *Marginal Consciousness*. On my account of "indexical" contents, contents of concrete acts are keyed to the relevant context of consciousness: see my 1989.

proposition I am thinking, and the content of my arithmetical thought makes no intentional appeal to these features of context.

VII. How Much do We See?

A recent experiment in cognitive science runs as follows. The subjects are instructed to watch a basketball game in progress (on videotape) and focus their attention on the number of times one team gains possession of the ball. During the course of play a man in a gorilla suit walks across the court in full view. Amazingly, the subjects do not notice the gorilla! What are we to make of such cases of "inattention blindness?"[7]

Traditional visual science assumed (with Mach and others) that the mind or brain constructs a model of the visual environment, presenting one with a detailed visual field of colors, shapes, and objects before one. Some skeptics argue from cases of inattention blindness that there is no such *visual field*, that our impression that we "see" such a rich visual display is illusory. Against this skepticism Alva Noë (2001) argues that we should analyze *visual awareness* as "a state of interactive engagement with the world, not a state of picture-making" (51). In the spirit of Merleau-Ponty's emphasis on the role of body in vision, Noë argues that perception is a process in which the whole person is engaged with objects so as to be ready to gather information from the environment as needed, without necessarily forming a detailed representation of the environment.

Noë is right in resisting the new skepticism and in underscoring the role of embodiment. Yet we need a structural analysis of attention and context to see exactly where the skeptic goes wrong as to what we see. Such an analysis we have developed through reflection on the Gurwitschean analysis of the field of consciousness.

Thus, we find, the *form* of visual awareness is not a simple matter. The ontological choice is not between putting "awareness" either inside or outside the mind (or its neural substrate). And the phenomenological choice is not between putting visual detail in a mental representation or putting it in the external environment. Intentionality in vision consists in a complex *intentional relation* involving (to use the first person): me, my visual experience, the meaning in the experience, the context of the experience, and the object of attentive consciousness. Granted the outlines of our neo-Gurwitschean analysis of context and context-awareness, we can put in place the several elements of visual intentionality without eliminating them or somehow losing track of them—as does the new skeptic.

---

[7] See discussions of this and other cases in Noë 2001 and O'Regan and Noë 2001.

Suppose I am the subject of inattention blindness. At a given moment in watching the basketball game, I have a conscious visual *experience* of seeing "that player with the ball." I am an embodied perceiving subject who has this experience in a particular context. My visual experience entertains an ideal meaning or *noema*, with the formal structure of attention. Thus, I see focally or *thematically* "that player with the ball." But there is more in my field of consciousness. I also see *relevantly* "his team member beside him" and "his opponent guarding him." And I also see *marginally* and non-relevantly "the empty chair behind him on the sideline." All these elements of content are inside my experience, in the sense that the noema is entertained or realized in the experience. As it happens, there is no item of content like "the gorilla figure walking behind him," not even in the marginal part of my field of consciousness. What is surprising is how strongly the *meaning* influences what is presented even marginally: the gorilla in the midst of play is so irrelevant to the meaning of the action that I do not even marginally notice the gorilla.

Now consider the *context* of my experience. The context or *Umwelt* of my experience includes me, my stream-of-consciousness (*cum* memory), my body (including the functioning of my eyes and optic neural system), the basketball game in progress (including the institutions of sport), the arena with seats and people watching the game, and so on. These entities of context lie outside my visual experience. Yet my experience *depends* ontologically on many of these things, often in ways I have no knowledge of. Some of these entities are accessible to vision, but others are known only by background understanding or mere habit. The content "that player with the ball" intentionally prescribes a certain object in the context of perception, where its intentional force is defined partly in terms of the formal attentional structure of the noema in my experience. As it happens, the gorilla is present in my *Umwelt* but no item of content in my experience registers its presence. What is surprising in this case is only the fact that a slight shift in attention, as would likely happen in an arena full of distractions from the game itself, would bring the gorilla into some kind of focus.

With a detailed analysis of the structures of context and context-awareness, then, we can deconstruct the new skeptic's dismissal of the visual field.[8]

---

[8] On Husserl's account, we add, the *horizon* of a perception includes not what is *actually* presented in the perceptual experience—whether focally, thematically, or marginally—but what is only *implicitly* presented: namely, those possibilities about the object and context that are allowed by the act's sense but not actually presented through that sense. (See Smith and McIntyre 1982.) When the skeptic denies all or most of the visual field, he is denying that this or that item in the environment is actually represented in the visual field. Fair enough, but beyond the field of actually visually represented entities in the context of perception lies the horizon of implicitly represented entities ("that" player's expected back side, the number on his jersey's back side, etc.). And still further out in the horizon lie "wild" possibilities like the gorilla's walking across the court. This possibility is not a "motivated" or relevant possibility, yet it is not completely ruled out by the

IX. Conclusion: The Formal Ontology of Intentionality

What we find above are the outlines of a formal ontology of intentionality-in-context: an analysis of the form of the intentional relation of act to object *in context*. The analysis features categorial forms of meaning (Thematic, Relevant, Marginal Meaning), categorial forms of context (Stream-of-Consciousness, Embodied-Existence, Perceptual-*Umwelt*), and specific forms of intentional relation wherein context plays an essential role invoked by noematic content.

These results will take their place in a larger phenomenological ontology that addresses the dynamical structure of conscious experience. As Gurwitsch and Husserl emphasized, my "field" of consciousness is a flux of experience amidst a flux of events in my *Umwelt*. Indeed, my moment of vision in seeing "that bird" is a part of my ongoing flux of consciousness-*cum*-action, wherein my conscious visual experience interacts with my embodied volitional action.[9] All this structure changes my simply seeing that bird into a part of my many-sided experience of living.[10]

References

Dreyfus, Hubert L. Editor (with Harrison Hall). 1982. *Husserl, Intentionality and Cognitive Science*. Cambridge, Massachusetts: MIT Press.

Embree, Lester. 2004. "The Three Species of Relevancy in Gurwitsch." In the present volume.

Gurwitsch, Aron. 1964. *The Field of Consciousness*. Pittsburgh: Duquesne University Press.

—. 1985. *Marginal Consciousness*. Edited by Lester Embree. Athens, Ohio: Ohio University Press.

Husserl, Edmund. 2001. *Logical Investigations*. Volumes I and II. Translated by J. N. Findlay; edited by Dermot Moran. London and New York: Routledge. Findlay trans-

---

noema of my seeing "that player with the ball."

9 Yoshimi (2003) emphasizes the dynamical structure of the "field of consciousness" analyzed by Gurwitsch. And Noë and Thompson (editors, 2003) explores the interaction between what we think of as simple vision and embodied action. Accordingly, my ongoing actions are typically part of the context of my vision, sometimes in the focus of attention and sometimes in the margin of attention.

10 Thanks to Lester Embree for organizing the conference, whose occasion prompted my return to Gurwitsch's work in a new context. Thanks to Jeff Yoshimi and Jason Ford for detailed comments on an earlier draft of this essay. And thanks to Lester Embree and all the participants for a variety of helpful comments and questions.

lation, first published in 1970. German original, first published in 1900-01, revised 1913, 1920.
—. 1969. *Ideas [pertaining to a Pure Phenomenology and a Phenomenological Philosophy, First Book]: General Introduction to Pure Phenomenology.* Translated by W. R. Boyce Gibson. London: George Allen & Unwin Ltd., and New York: Humanities Press, Inc. First English edition, 1931; German original, first published in 1913. Called *Ideas I.*
Kaplan, David. 1989. "Demonstratives." In Almog, Joseph, John Perry, and Howard Wettstein, editors. *Themes From Kaplan.* Oxford and New York: Oxford University Press. 481-563.
Noë, Alva. "Experience and the Active Mind." *Synthese* 129 1 (October 2001) 41-60.
Noë, Alva, and Evfan Thompson. 2002. Editors. *Vision and Mind: Selected Readings in the Philosophy of Perception.* Cambridge, Massachusetts: MIT Press.
O'Regan, J. Kevin, and Alva Noë. 2001. "What it is Like to See: A Sensorimotor Theory of Perceptual Experience." *Synthese* 129 1 (October) 79-103.
Perry, John. 2001. *Knowledge, Possibility and Consciousness.* Cambridge, Massachusetts: MIT Press.
Petitot, Jean, Francisco J. Varela, Bernard Pachoud, and Jean-Michel Roy, editors. 1999. *Naturalizing Phenomenology: Issues in Contemporary Phenmenology and Cognitive Science.* Stanford, California: Stanford University Press in collaboration with Cambridge University Press, Cambridge and New York.
Searle, John. 1983. *Intentionality.* Cambridge and New York: Cambridge University Press.
Smith, Barry, and David Woodruff Smith, editors. 1995. *The Cambridge Companion to Husserl.* Cambridge and New York: Cambridge University Press.
Smith, David Woodruff. 1970. *Intentionality, Noemata, and Individuation: the Role of Individuation in Husserl"s Theory of Intentionality.* Stanford, California: doctoral dissertation, Stanford University.
—. 1982. "Husserl on Demonstrative Reference and Perception." In Hubert L. Dreyfus, editor, *Husserl, Intentionality and Cognitive Science.* Cambridge, Massachusetts: MIT Press. 193-213.
—. 1989. *The Circle of Acquaintance: Perception, Consciousness, and Empathy.* Dordrecht and Boston: Kluwer Academic Publishers.
—. 1995. "Mind and Body." In Smith and Smith, editors. *The Cambridge Companion to Husserl.* Cambridge and New York: Cambridge University Press.
—. 1999. "Intentionality Naturalized?" In Petitot *et al.*, editors, *Naturalizing Phenomenology: Issues in Contemporary Phenmenology and Cognitive Science.* Stanford, California: Stanford University Press in collaboration with Cambridge University Press, Cambridge and New York.
Smith, David Woodruff, and Ronald McIntyre. 1982. *Husserl and Intentionality: A Study of Mind, Meaning, and Language.* Dordrecth and Boston: D. Reidel Publishing Company, now Kluwer Academic Publishers.

Chapter 10

The Field of Consciousness as a Living System: Toward a Naturalized Phenomenology of Cognition

Shinya NOE
Tohoku Institute of Technology

Abstract

Ever since cognitive science emerged as a "science of consciousness," we have been confronted with a challenging question concerning the self-understanding of phenomenology, i.e., the question of how we should think of the relationship between phenomenology and cognitive science. In considering this question, one possible position that one could take would be to clarify the proper character of the "philosophical question" in contrast to that of the "scientific question" concerning the issues of knowledge. Then it could be argued, for example, that cognitive science deals with "cognitive process as a fact," whereas phenomenology deals with "the conditions of possibility of cognitive process." But such an argument tends to lead to a contrast between the scientific approach in general (including cognitive science) and the philosophical approach in general, and to pass over the issue proper to cognitive science, especially that of its methodology. In this essay I will not take this position. Firstly, I would like to clarify the philosophical implications of the emergence of cognitive science in connection with phenomenology. In my opinion we could gain a clear picture of the issue if we take the concept of "systems" into consideration as the medium between phenomenology and cognitive science.

I. Logical, Physical and Mental Systems

The first clue to an understanding of the relationship between phenomenology and cognitive science lies in reconsidering the relationship between Husserl and Frege. In particular, it is important to shed some light upon their views on "the mental." In his celebrated paper "On *Sinn* and *Bedeutung*" (1892) Frege introduces the threefold distinction between *Sinn*, *Bedeutung*, and *Vorstellung*. He uses the following analogy in order to clarify these relationships: "Somebody

L. Embree (ed.), Gurwitsch's Relevancy for Cognitive Science, 187-204.

observes the Moon through a telescope."[1] He compares the Moon itself to the *Bedeutung*; it is the object of the observation. The real image projected by the object glass in the interior of the telescope is compared to the *Sinn*, and the retinal image of the observer is compared to the *Vorstellung*. The real image is objective, inasmuch as it can be used by several observers while the retinal image is subjective, inasmuch as "each one would have his own retinal image."[2] He emphasizes the objectivity of *Sinn* in contrast with the subjectivity of *Vorstellung*, arguing that the common property of many people, and so is not a part or a mode of the individual mind"[3] and, in the end, excludes the *Vorstellung* from the region of logic. He concludes his analysis with a wholly negative remark: "In what follows there will be no further discussion of *Vorstellungen* and *Anschauungen*."[4] In this way, in Frege' s case, the mental is regarded as essentially private and subjective, and it stands in striking contrast with the logical regarded as public and objective.

On the other hand, for Husserl the mental stands in contrast not only with the logical in Fregean sense but also with "the real." He learned through Brentano that the intentional relations to objects that are distinctive of mental phenomena are by no means "real," so that the mental is not reducible to the physical as the real. Therefore, in Husserl's case, the subject matter of science can be classified into three main regions—i.e., the logical, the physical, and the mental. In the terminology of systems theory, three kinds of systems come into question: logical systems as the ideal (systems I), physical systems as the real (systems II), and mental systems as the transcendental (systems III)

To use the Kantian terminology, systems I are formal and a priori, systems II are material and a posteriori, and systems III are material and a priori. When Frege insisted on the autonomy of the logical vis-à-vis the mental, he thought that the mental totally belongs to the real, so that in Frege's case the subject matter of science can be classified into two kinds of systems—i.e., systems I (the subject matter of logic and mathematics) and systems II (the subject matter of empirical science), and there is no place for systems III. This either-or stance is widely accepted by analytic philosophers who inherit the legacy of Frege. In most cases they thought that the mental is reducible ultimately to the physical.

On the other hand, in the first volume of his *Logical Investigations* (1900), Husserl moved towards an anti-psychologism and claimed that the laws of logic are independent of the laws of psychology. Thereby he shared the historical point of departure with the analytic tradition emanating from Frege. In fact, Husserl had a thorough knowledge of the logical literature of his day including Frege' s work and was well versed in the techniques of symbolic logic. Accord-

---

[1]M. Beaney (ed.), *The Frege Reader* (Oxford: Blackwell, 1997), p. 155.
[2] Ibid.
[3] Ibid.
[4] Ibid.

ing to Marvin Farber, "Husserl might well have become a leader in the development of symbolic logic."[5] But the subsequent development of Husserl' s phenomenology after the second volume of his *Logical Investigations* (1901) was oriented towards the analysis of consciousness in striking contrast to the logical analysis of language. They were analytic philosophers that constructed symbolic logic and they considered phenomenology to be "metaphysical." This has been regarded as a paradox so far. But, as Elmar Holenstein suggests, Husserl' s critique of psychologism had a "purifying function."[6] After clarifying the autonomy of the laws of logic vis-à-vis the laws of psychology, Husserl tried to make it clear that the laws of psychology similarly have a certain autonomy vis-à-vis the real laws of biology and physics and, moreover, that the laws of psychology have a certain autonomy vis-à-vis the ideal laws of logic.

For Frege the question is to separate the logical from the real in order to purify the region of the logical and the most important thing is to reach the logical objects as *Bedeutung.* According to him, *Bedeutung* is always given through the medium of *Sinn.* But for him even *Sinn* is no more than something like a ladder. We must throw it away after we have climbed up it to reach the *Bedeutung,* although it is a necessary medium for *Bedeutung.*

On the other hand, for Husserl the question is to separate the mental from both the logical and the real in order to purify the region of the mental. Although he acknowledges the autonomy of ideal beings like logical and mathematical objects, he thinks of them not as objects wholly independent of consciousness, but as intentional correlates of the acts of consciousness such as judgment or representation and seeks to elucidate their essence in the intentional correlation with consciousness. He faces the following question: how can objectivity be constituted through the subjective performances of the individual ego?—In other words, how can two kinds of systems that are autonomous and independent of each other, i.e., the logical systems and the mental systems relate?

Husserl acknowledges the existence of "material a priori" and advocates "intuition of essence" as the method of analysis on the level where material a priori truth is possible. He seeks to purify the mental from "psychological phenomena as real occurrences" and recaptures them as "irreal" and advocates "transcendental reduction" as the method of such purification. He says that "the phenomena of transcendental phenomenology will become characterized as irreal" and that transcendental reductions "purify" mental phenomena from what confers reality on them and, with that, their place in the real "world."[7] Moreover, Husserl insists that "pure or transcendental phenomenology will become established, not

[5] M. Farber, *The Foundation of Phenomenology* (Cambridge, MA: Harvard Univ. Press, 1943), p. 89.

[6] E. Holenstein, *Menschliches Selbstverständnis* (Frankfurt am Main: Suhrkamp, 1985), p. 12.

[7] E. Husserl, *Ideas I*, trans. F. Kersten (Dordrecht: Martinus Nijhoff, 1982), p. xx.

as a science of matters of fact, but as a science of essence (as an "eidetic" science)" and that eidetic reduction leads over from the *de facto* mental phenomena to the pure "essence."[8] An essence gives rise to a "material a priori" truth. Husserl's search for essence is parallel to Kant's demand for "synthetic a priori" truth. An essential truth is neither analytically ("formally") true nor is it merely materially ("empirically") true.

Analytic philosophers have been attacking the very idea of essential truth and the evidence claimed for it since the time of Logical Positivism. Moritz Schlick, for example, approved Kant's identification of the a priori with the formal while he blamed phenomenologists for fixing the boundaries of a priori judgments astonishingly wider than Kant. He asserted that only analytic propositions are a priori, so that there is no material a priori. He wrote:

> A philosopher who believes in a material a priori, and would like to clarify its possibility, would have, as far as I can see, no other way out than to carry over the Kantian theory from the form to the content of experience. He would have to assume that not only the form of our cognitions, but also their matter, springs from the knowing consciousness—for only thus could a priori propositions relating to them be made intelligible. This would amount to a subjective Idealism of the Fichtean variety; one would find oneself entangled in a weird metaphysics.[9]

In this way analytic philosophers have been thinking that only two levels of analysis are possible in science: that is, the level of "formal and a priori" analysis of logic and mathematics and that of "material and a posteriori" analysis of empirical science, and that acceptance of the third level of analysis is identified with belief in "metaphysics."

But cognitive science showed that the third level of analysis is possible in science. Cognitive science is the study of cognitive activity fully abstracted in principle from biological foundations. Zenon Pylyshyn says that Alan Turing's work can be seen as the first study of such kind. According to Pylyshyn, "it represents the emergence of a new level of analysis, independent of physics yet mechanistic in spirit. It makes possible a science of structure and function divorced from material substance."[10] This new science is in no way "metaphysics," and it is phenomenology that represents the "condition of possibility" of the new science by making it clear that there is a level of "structural and functional" analysis.

---

[8] E. Husserl, op. cit., p. xx.
[9] M. Schlick, *Philosophical Papers*, vol. II (Dordrecht: D. Reidel, 1979), p. 164.
[10] Z. Pylyshyn, "Complexity and the Study of Artificial and Human Intelligence," in J. Haugeland (ed.), *Mind Design* (Cambridge, MA: MIT Press, 1981), p. 68.

II. Structural Analysis of Systems

The specific character of phenomenological analysis may be clearly described in the connection with the analysis of structural linguistics of Roman Jakobson, who was deeply influenced by Husserl in his linguistic methodology before he knew Saussure's new language theory, as can be seen in the following quotation:

> At the beginning of our century the thought of Husserl, developed in the second volume of his *Logische Untersuchungen* and particularly in the chapter where he treats "the difference between independent and dependent meaning and the idea of pure grammar," became a powerful factor for the first steps of structural linguistics by superimposing "the idea of a general and *a priori* grammar" on "the exclusively empiric" grammar which at that time was the only one accepted.[11]

Moreover, as Holenstein points out, Jakobson found in Husserl's work "the first systematic formulation of the general laws that are operative for a structural unit,"[12] and regarded the third Logical Investigation, entitled "On the Theory of Wholes and Parts," as a kind of "fundamental view of structuralism."[13] Jakobson tried to reform the orthodox historical linguistics of his day in the same way as Husserl tried to reform elementalistic psychology, i.e., Jakobson treated language as a "system," a unified whole divorced from outer factors and tried to elucidate the laws that are constitutive of it (phonological laws would be good examples). He brought forward the key idea that is characteristic of the new science as follows:

> Any set of phenomena examined by contemporary science is treated not as a mechanical agglomeration but as a structural whole, and the basic task is to reveal the inner, whether static or developmental, laws of this system. What appears to be the focus of scientific preoccupations is no longer the outer stimulus, but the internal premises of the development; now the me-

---

[11] R. Jacobson, *Main Trends in the Science of Language* (New York: Harper & Row, 1974), p. 13.
[12] E. Holenstein, *Roman Jacobson's Approach to Language* (Bloomington, IN: Indiana University Press, 1976), p.2.
[13] Ibid.

> chanical conception of processes yields to the question of their functions.[14]

The "structural and functional" analyses of Husserl and Jakobson are designed to counteract the mechanistic-causal explanation prevalent toward the end of the nineteenth century. Jakobson opposed orthodox methodology in historical linguistics in the same way as Husserl opposed elementalistic methodology in psychology. Their approach to language and consciousness is symbolic of the historic change of course in scientific thinking.

Systems theory is a later product of the scientific developments in the 1930s represented by such new sciences as cognitive science emanated from Turing's theory of computation, Jacobson's structural linguistics, and Gestalt psychology. Their methodology is characterized by the revolt against Newtonian view of classical sciences (mechanism, reductionism, and causal thinking). It is interesting that what Ervin Laszlo, a pioneer of systems philosophy, talks about the systems view in general corresponds completely with what Jacobson talks about the methodology of the new science: "The systems view always treats systems as integrated wholes of their subsidiary components and never as the mechanistic aggregate of parts in isolable causal relations."[15]

In this way, the movements of reforming science in 1930s and systems theory as a later product of such movements have a close affinity to Husserl's static phenomenology, which is characterized by transcendental and eidetic reduction. In the terminology of systems theory, the focus of their scientific preoccupations is "structure" of systems. As we have seen above, the term "system" means a unified whole divorced from outer factors. On the other hand, the term "structure" means a system viewed from the standpoint of invariant relations of their constituents.

Aron Gurwitsch's examination of the articulation of consciousness is most typical of the "structural analysis" of this kind. It was a natural consequence for Gurwitsch who noticed that there is something in Gestalt psychology that is convergent with the insights of Husserl's transcendental phenomenology. According to Merleau-Ponty—who was deeply influenced by Gurwitsch—when Gestalt psychology made the first attack on elementalistic psychology, it adopted a "kind of reflection the theory of which is furnished by Husserl's phenomenology."[16] As for the historical relation between phenomenology and Gestalt psychology, Merleau-Ponty pointed out that Kurt Koffka, one of the three principal members of the Berlin school, had expressly recognized his debt to phenomenology. He also referred to a "diffuse influence that was not always

---

[14] R. Jacobson, op. cit., p. 11.

[15] E. Lazlo, *The Systems View of the World* (New York: George Braziller, 1972), pp. 14 f.

[16] M. Merleau-Ponty, *Phenomenology of Perception*, trans. C. Smith (New York: Humanities, 1962), p. 50 n.

intended by Husserl or recognized by those who experienced it"[17] and remarked: "All that was done in Germany from 1915 to 1920 under the direct or indirect influence of phenomenology was by no means accomplished exclusively in the courses of Husserl."[18]

Gestalt psychologists maintained that psychological phenomena could not be understood as associations between stimuli and responses. For example, according to their interpretation, perception is not a passive receiving of incoming stimuli, but an active process of re-structuring or reorganizing of the coherent whole. From this point of view, Gestalt psychologists criticized the constancy principle of elementalistic psychology about the strict correlation between psychological and physical events. Gurwitsch interpreted transcendental reduction as primarily a rejection of the constancy principle. For him it means to divorce psychological events from physical events as outer factors, i.e., to treat consciousness as a "system."

According to Gurwitsch, the field of consciousness is differentiated into three organizational domains with their own respective inherent principle of organization. The first domain is the theme, which "engrosses the mind of the experiencing subject" and occupies the "focus of attention" (controlled by the Gestalt-coherence). The second domain is the thematic field defined as everything that is "co-present" with the theme and experienced as materially relevant to the theme (controlled by the Gestalt principle of relevance). The third domain is the margin, which encompasses the data that are co-present with the theme without having any material relevancy to it. Gurwitsch claims that this "theme-thematic-field-structure is a formal invariant of organization having universal significance. The structure is realized at every moment of conscious life."[19]

III. Autopoietic Systems

Gurwitsch's analysis of consciousness and Jacobson' s analysis of language correspond to the development of systems theory till the 1970s, in which various theoretical efforts were focused on elucidating "whole-part" relation or "founding" relation of the constituents of systems, all aimed at describing the order of systems as "structure." In systems theory since the 1970s, however, a remarkable conversion occurred with the background of discoveries of self-organizing phenomena of systems on various levels. The becoming or developing of the order of systems has become a focus of attention instead of "structure." This

---

[17] M. Merleau-Ponty, *The Primacy of Perception*, trans. J. Edie (Evanston, IL: Northwestern University Press, 1964), p.47.
[18] Ibid.
[19] A. Gurwitsch, *The Field of Consciousness* (Pittsburgh, PA: Duquesne University Press, 1964), p. 379.

new trend in systems theory is represented by the theories of autonomous, self-referring, and self-constructing closed systems—in short, autopoietic systems of Francisco Varela, Humberto Maturana, Niklas Luhmann, etc. According to Maturana, autopoietic systems are "systems that are defined as unities as networks of productions of components that recursively, through their interactions, generate and realize the network that produces them and constitute, in the space in which they exist, the boundaries of the network as components that participate in the realization of the network."[20]

Varela and Luhmann acknowledged that they were deeply influenced by Husserl in their methodology. This suggests a possibility of integrating Husserl's transcendental phenomenology with the theory of autopoietic systems. In fact there are noteworthy internal relationships between Husserl's thought and the theory of autopoietic systems both in respect of historical contexts and in respect of theoretical results obtained from them. Historically stated, Husserl's turn from static to genetic analysis of consciousness corresponds to the development from structural to autopoietic standpoint in systems theory. In particular, Husserl's analyses of passive constitution pursued in the 1920s under the title of "analyses of passive synthesis" could be interpreted as an anticipatory description of self-organizing process of consciousness as an autopoietic system.[21]

According to Luhmann, consciousness is a self-referential autopoietic system, because consciousness is a living system and life is defined as autopoiesis. However, we immediately get into trouble in precisely defining what the components of consciousness are. Luhmann presented the thesis that *Gedanken* or *Vorstellungen* are the components of consciousness in his article "Autopoiesis of Consciousness" in 1985. But his thesis is misleading, because we can still find the elementalistic bias of nineteenth-century empirical psychology there. We should remember here that Husserl opposed elementalistic methodology in nineteenth-century psychology. Following Brentano, he considered mental phenomena as those phenomena that intentionally contain objects within themselves and found inseparable correlations between acts of intention (noeses) and objects of such acts (noemata)—for example, as Brentano said, no hearing without something heard, no believing without something believed, etc. We should think of the components of consciousness, following Husserl's insights, noetico-noematic correlations.

As Klaus Held says, consciousness is not like a container that can be filled with any kind of thing. Consciousness consists of various acts (noeses) whose character is in each case determined by the kind of corresponding object, and the object can appear to consciousness only in the corresponding manner of given-

---

[20] H. Maturana, "Autopoesis," in M. Zeleny (ed.), *Autopoesis: A Theory of Living Organization* (New York: Elsevier, 1981), p. 21.

[21] Edmund Husserl, *Analyses concerning Passive and Active Synthesis*, translated by Anthony J. Steinbock (Dordrecht: Kluwer Academic Publishers, 2001).

ness. Here Husserl sees the "universal a priori of correlation between experienced object and manners of givenness."[22] That is to say, when we perceive an object, it is given as what is perceived in consequence of the productivity of consciousness (the productivity is not necessarily "active," it can be also "passive"); when we remember an object, it is given as what is remembered in like manner and we never confuse what is perceived with what is remembered. What follows from this is that consciousness reveals three significant characteristics.

Firstly, consciousness makes application of its operations to the results of its own operations. In this way, it constitutes the components of the system (noema, that is, an object captured in its manner of givenness) in a self-referential way. Consciousness is a network of intentional correlations of noesis and noema. It is a self-contained unity whose only reference is to consciousness itself.

Secondly, according to Husserl after his "transcendental turn," consciousness is not only "consciousness of something," but also, conversely, "everything" is only in relation to consciousness, i.e., relative to it. Consciousness is "absolute" in the sense that it needs no other thing than consciousness itself in order to be, therefore completely autonomous.

Thirdly, consciousness is an operationally closed system, which comprises "everything." It produces and reproduces elements (and boundaries) in a recursive way. It has no outside, or, in other words, an outside is precisely nonsense.

To summarize the major characteristics of consciousness, self-reference, autonomy, and the absence of outside are of most significance. Transcendental consciousness is nothing else but the consciousness that bears these structural characteristics. Here we can see a noteworthy structural isomorphism between transcendental consciousness and autopoietic systems. The elements of transcendental consciousness are self-referentially constituted noetico-noematic correlations. They form a network by linking up together. Noetico-noematic correlations are recursively produced and reproduced by a network of noetico-noematic correlations.

If transcendental consciousness is an autopoietic system, this seems to suggest a revised concept of intentionality. The concept of intentionality, often defined as "directedness to objects," implicitly presupposes the distinction between "inside" and "outside" of consciousness. It is apparently incompatible with the transcendentality of consciousness, because transcendental consciousness has no "outside," therefore no "inside" as well. Should we, then, abandon the concept of intentionality? If we do so, consciousness loses its dynamism. The only way of solving the problem is to interpret intentionality as a recursive self-relation of consciousness. Only when we consider consciousness in this way, we can say that consciousness is a living system, because self-reference is the most inherent principle of life.

---

[22] E. Husserl, *The Crisis of European Sciences and Transcendental Phenomenology*, trans. D. Carr (Evanston, IL: Northwestern University Press, 1981), pp. 40-42.

At this point there is abundant reason why Varela, one of the most eminent cognitive scientists and most staunch defenders of the theory of autopoietic systems, took a serious interest in phenomenology. "Naturalized phenomenology" is the name for the final stage that he reached after a long career as a cognitive scientist. In his case "naturalization" does not mean "naturalism." He sought to integrate the insight of phenomenology with the knowledge of cognitive science, not by "naturalizing" consciousness, but by radicalizing the transcendentality of consciousness, through the mediation of the theory of autopoietic systems.

The theory of autopoiesis sees the autonomy of consciousness in the self-reproduction of a network. But at the same time consciousness has a relation to environment, and the most significant "environment" for it is the nervous system of the brain, which itself is another autonomous network. The theory of autopoiesis understands the relation of consciousness to the nervous system as interference between the two autonomous networks. It offers a chance of getting around the falsely posed alternative between reductionism and dualism. In brief, consciousness as an autopoietic system distinguishes itself from the nervous system, another autopoietic system, by constituting components in a self-referential way and linking them together in a hypercycle. Figure 1 below indicates the structure of transcendental consciousness as a total system, in which self-referentially constituted components (small cycles, noetico-noematic correlations) are hypercyclically coupled.

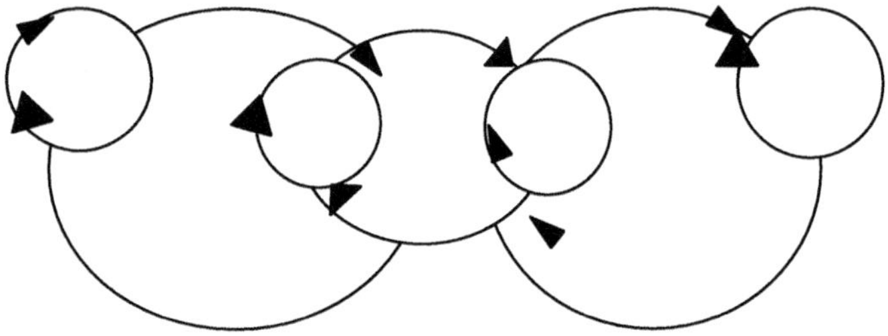

Fig. 1. Transcendental consciousness as an autopoietic system

As we have said already, transcendental consciousness comprises everything; it comprises even the system called "ego." Ego is in fact a second-order autopoietic system; it arises as a self-referentially constituted component (functional system) from transcendental consciousness, the first-order autopoietic system. Transcendental consciousness is absolutely unique, whereas ego can be plural, as in the case of dual personality; it can be disappeared, as the Buddhist doctrines of no-ego indicate. Ego is thus a partially autonomous system and self-organizing. Self-organization refers to the ability of a system spontaneously to produce an autonomous order. Order is not imposed from the outside, but is produced internally through the interplay of the components of the system.

According to Varela, cognitive science has been developing in three phases:

Phase 1: Cognitivist (computationalist) approach
Phase 2: Connectionist approach
Phase 3: Enactive approach

In order to clarify the significance of enactive approach of Varela, I begin with a presentation of the historical background of computationalism and connectionism.

Cognitive science theorists have been seeking analogues to the mind in order to account for the mind's activities. In cognitive science, two analogues have provided the bases for modeling of cognitive performance. On the one hand, the symbol processing computer became a plausible analogue to the mind and numerous cognitive science theorists have been attracted to the proposal that the mind itself is a symbol processing device. The other analogue was the brain. The view that the brain consisted of a network of simple processing units which excited and inhibited each other became popular. The second approach came to be known by a number of names, including "connectionism," "neural networks," and "parallel distributed processing." Over the last decade, this emerging paradigm of computational modeling has steadily gained the respect of many theorists as a radically new and promising approach to cognitive science.

On the other hand, like cognitive science theorists, Husserl investigated the mind's activities, but of course not as empirical science based on the "naturalization" of consciousness, but as transcendental phenomenology, i.e., by purely nonempirical methods. And, moreover, he, at least explicitly, did not have any analogue to the mind. Nevertheless, there is a noteworthy parallelism between the connectionist approach in cognitive science and the analyses of passive synthesis in Husserl's genetic phenomenology.

IV. Computationalism

Cognitive science emerged as the computer-simulation study of human cognitive processes in the late 1950s. In 1958 Allen Newell and Herbert Simon provided a computer program named LT (Logic Theorist) which demonstrates the theorems of propositional logic. They presented the program itself as a psychological theory of human thought processes. This was an epoch-making event which drove psychology toward a paradigm shift.

Cognitive scientists regard human cognitive processes as the output of an information processing system with a certain structure, and describe the structure and function of cognitive capacities in terms of empirically testable expressions. The typical methodology adopted by cognitive scientists is as follows: cognitive scientists formulate a hypothesis concerning the cognitive processes under examination as a computer program, and they test the hypothesis by comparing it with the operations which the computer performs by means of the program. In this way the methodology of cognitive science is similar to the classical methodology adopted in the empirical sciences: that is, formulation of a hypothesis in terms of formal expressions, and verification of the hypothesis by experiment. This is why cognitive science is generally regarded as an "empirical science" of the mind.

The methodology of cognitive science is formed under the influence of computer science. Its basic presupposition is that both computers and humans are symbol-processing systems, which is most clearly expressed in Newell and Simon's hypothesis of a physical symbol system, "the necessary and sufficient condition for a physical system to exhibit general intelligent action is that it be a physical symbol system."[23] A physical symbol system, according to Newell and Simon, is any member of a general class of physically realizable systems meeting the following conditions:

> (1) It contains a set of symbols, which are physical patterns that can be strung together to yield a structure (or expression).
> (2) It contains a multitude of such symbol structures and a set of processes that operate on them.
> (3) It is located in a wider world of real objects and may be related to that world by designation or interpretation.[24]

Newell and Simon claim that any generally intelligent physical system will be a physical symbol system (the necessity claim) and that any physical symbol system "can be organized further to exhibit general intelligent action" (the sufficiency claim). And "general intelligent action," on Newell and Simon's gloss, implies "the same scope of intelligence seen in human action."[25]

Newell and Simon share the view that the essence of intelligence lies in computation, and intelligence could be realized by a universal machine capable of doing any well-specified computation, i.e., by a physical symbol system. Their

---

[23] A. Newell and H. Simon, "Computer Science as Empirical Inquiry," in J. Haugeland (ed.), *Mind Design* (Cambridge, MA: MIT Press, 1981), pp. 40-42.
[24] Ibid.
[25] Ibid.

stance is known as computationalism. According to Varela, the computationalist (in his terminology, the cognitivist) research program can be summarized as answers to the following fundamental questions:

> Question 1: What is cognition?
> Answer: Information processing as symbolic computation—rule-based manipulation of symbols.
> Question 2: How does it work?
> Answer: Through any device that can support and manipulate discrete functional elements—the symbols. The system interacts only with the form of the symbols (their physical attributes), not their meaning.
> Question 3: How do I know when a cognitive system is functioning adequately?
> Answer: When the symbols appropriately represent some aspect of the real world, and the information processing leads to a successful solution of the problem given to the system.[26]

According to computationalism, intelligent action is a symbolic computation, i.e., formal manipulation of symbolic representations according to algorithm (a set of processes that operate on symbol structures). Such intelligent actions as calculation and logical inference are typical symbolic computations. It is clear that any von Neumann computer of today constitutes physical symbol system, which performs symbolic computation as technological reproduction of conscious thought process of human agents.

V. Connectionism

Von Neumann computers exceed human agents in the ability of calculation and logical inference, whereas they do not excel in understanding by analogy, intuitive judgment, pattern perception etc., which human agents perform in a natural way. They are too structured and inflexible to be able to cope with all the variants of a situation. Yet human agents are sensible enough to take such variants in stride and to adjust themselves flexibly even to unknown conditions. Moreover, their actions are modifiable as a function of experience, i.e. they exhibit learning.

---

[26] F. Varela, E. Thompson, and E. Rosch, *The Embodied Mind* (Cambridge, MA: MIT Press, 1999), pp. 42f.

Where, then, do such abilities and flexibilities proper to human agents come from? The connectionist approach is an attempt to answer this question. It is presented as an alternative to the classical approach of computationalism. The conscious thought process that is linguistically realized is, in the terminology of information technology, serial, which means that information is processed sequentially. On the connectionist view on human cognition, such a serial information-processing is based on parallel information-processing, where a number of processes are carried out simultaneously. Human parallel information-processing is supported by the dynamics of parallel interactions in the neural network of the brain. In connectionist architectures, it is replaced by PDP (Parallel Distributed Processing) systems. Paul Smolensky sums up the point:

> Connectionist systems are large networks of extremely simple processors, massively interconnected and running in parallel. Each processor has a numerical activation value which it communicates to other processors along connections of varying strengths. The activation value for each processor constantly changes in response to the activity of the processors to which it is connected. The values of some of the processors form the input to the system, and the values of other processors form the output. The connections between the processors determine how input is transformed to output.[27]

Smolensky describes PDP systems as working in what he calls the subsymbolic paradigm. What is important is that "in connectionist systems, knowledge is encoded not in symbolic structures but rather in the pattern of numerical strengths of the connections between processors."[28] In the subsymbolic paradigm, human thought process is not modeled by the manipulation of symbols. Rather the greater part of our thought process is subsymbolic or pre-linguistic. Varela formulates the connectionist approach in cognitive science by answering the same questions previously posed to computationalism:

> Question 1: What is cognition?
> Answer: The emergence of global states in a network of simple components.
> Question 2: How does it work?
> Answer: Through local rules for individual operation and rules for changes in the connectivity among the elements.

---

[27] F. Smolensky, "Connectionist AI, Symbolic AI, and the Brain," *Artificial Intelligence Review*, Vol. 1 (1987), p. 95.
[28] Ibid.

> Question 3: How do I know when a cognitive system is functioning adequately?
> Answer: When the emergent properties (and resulting structure) can be seen to correspond to a specific cognitive capacity—a successful solution to a required task.[29]

What, then, are the implications of connectionism for phenomenology? First of all, it is important to notice that the connectionist viewpoint is effective in criticizing the methodology of analytic philosophy, which has been characterized in opposition to that of phenomenology.

Since the "linguistic turn," analytic philosophy has been exclusively characterized as linguistic philosophy: that is, "the view that philosophical problems are problems which may be solved (or dissolved) either by reforming language, or by understanding more about the language we presently use."[30] The main reason why analytic philosophers have chosen language as a substantial clue for solving philosophical problems is that linguistic expressions are "observable things" and conform to the empiricist standard that "only sense makes sense."[31] According to Quine, the policy of talking about linguistic expressions where possible instead of ideas in epistemology entered Modern Empiricism in the late Eighteenth Century.[32] To pursue this policy means, on the other hand, to take minute precautions against talking about phenomena which are publicly unobservable, such as ideas or consciousness, since statements about publicly unobservable things lead to "metaphysics."

Such an attitude is essentially common to behavioristic psychology, so that it is not surprising that linguistic philosophers found themselves in a certain degree of sympathy with psychological behaviorism. Just as behavioristic psychologists wanted to reduce statements about such unobservables as a person's sensations or emotions to statements about such public observables as bodily behaviors, so linguistic philosophers were committed to the identification of thought and performance of linguistic ability. According to Michael Dummett, for example, "the only proper method for analyzing thought consists in the analysis of language."[33] Linguistic philosophers maintain that "thought without language" does not exist, but in reality the linguistic thought process is nothing but a secondary translation of the final stage of the pre-linguistic thought process. As Marvin Minsky says, "our thinking-in-words reveals only a fragment of

---

[29] F. Varela et al., op. cit., p. 99.
[30] R. Rorty, "Introduction," in R. Rorty (ed.), *The Linguistic Turn* (Chicago, IL: The University of Chicago Press, 1981), p. 3.
[31] W. V. Quine, *Theories and Things* (Cambridge, MA: Harvard University Press, 1981), p. 68.
[32] Ibid., p. 67.
[33] M. Dummet, *Truth and Other Enigmas* (Cambridge, MA: Harvard University Press, 1978), p. 458.

the mind' s activity"[34] and does not control directly our behavior in everyday life.

The interest of connectionists focuses not on conscious information-processing as the manipulation of explicit, symbolic representations (predicative process in Husserl's terminology), but on passive information-processing as the formation and transformation of subsymbolic representations (prepredicative process in Husserl's terminology). Passive process such as this, which is self-organizing and unconsciously realized, is just what makes it possible for us to behave appropriately in everyday life, according to our respective situations, contexts, goals, intentions, etc., and, in particular, to adjust ourselves flexibly even to unknown conditions. In my opinion what Husserl did under the title of "analyses of passive synthesis" is a descriptive anticipation of what the connectionists of our own day are doing.

VI. Enaction

The present debate in the theory of cognition is dominated by an unfruitful schism. On the one hand, idealistic theories are concerned with the autonomy of consciousness as a cognitive system, running the risk of missing the relation of consciousness to environment. On the other hand, the approaches of cognitive science (computationalism and connectionism), which are characterized by input-output paradigm of system-environment relations, analyze all sorts of interactions between cognitive system and environment, but have developed no conceptual tools to do justice to the autonomy of consciousness.

Taking visual perception as an example, Varela formulates the way of raising questions in cognitive science up to this time as the question: "Which came first, the world or the image?" The answer of most vision research—both computationalist and connectionist—is the *chicken position*:

> *Chicken position:* The world out there has pregiven properties. These exist prior to the image that is cast on the cognitive system, whose task is to recover them appropriately (whether through symbols or global subsymbolic states).[35]

We tend to think that the only alternative is the *egg position:*

---

[34] M. Minsky, *The Society of Mind* (New York: Simon and Schuster, 1985), p. 196.
[35] F. Varela et al., op. cit., p. 172.

> *Egg position:* The cognitive system projects its own world, and the apparent reality of this world is merely a reflection of internal laws of the system.[36]

Varela suggests a middle way between these chicken and egg positions. His theory of autopoiesis, which is characterized by closure paradigm of recursive production of elements (and boundaries), offers a chance of negotiating a middle path between the Scylla of cognition as the projection of a inner world (subjective idealism) and the Charybdis of cognition as the environmentally conditioned information-processing (scientific realism). His intention is to bypass these two extremes by studying cognition not as recovery or projection but as "enaction" (embodied action). Varela formulates the enactive approach in cognitive science by answering the same questions posed to computationalism and connectionism:

> Question 1: What is cognition?
> Answer: Enaction. A history of structural coupling that brings forth a world.
> Question 2: How does it work?
> Answer: Through a network consisting of multiple levels of interconnected, sensorimotor subnetworks.
> Question 3: How do I know when a cognitive system is functioning adequately?
> Answer: When it becomes part of an ongoing existing world (as the young of every species do) or shapes a new one (as happens in evolutionary history).[37]

Computationalism and connectionism both presuppose the existence of external world surrounding cognitive system and take representation (internal model of external world) as their central notion. They see cognitive system from the standpoint of "outer observer" and describe the interactions between cognitive system and external world according to input-output paradigm. For Varela, on the contrary, the most important thing is to carry out a radical conversion from the standpoint of an "outer observer" to that of the "system itself." To understand the standpoint of the "system itself," Maturana's metaphor is of great use:

> What occurs in a living system is analogous to what occurs in an instrumental flight where the pilot does not have access to the outside world and must function only as a controller of the values shown in his flight instruments. His task is to secure a

[36] Ibid.
[37] Ibid., p. 206 f.

> path of variations in the readings of his instruments, either according to a prescribed plan, or to one that becomes specified by these readings. When the pilot steps out of the plane he is bewildered by the congratulations of his friends on account of the perfect flight and landing that he performed in absolute darkness. He is perplexed because to his knowledge all that he did at any moment was to maintain the readings of his instruments within certain specified limits, a task which is in no way represented by the description that his friends (observers) make of his conduct.[38]

For Varela, Husserl's "transcendental turn" corresponds with the conversion from the standpoint of an "outer observer" to the standpoint of the "system itself." Both are the approaches to do justice to consciousness as a living system that is entirely self-referential and has no "outside." Seen from the standpoint of the "system itself," there is neither input nor output, because there is no "outside." The world is not "out there" independent of our cognitive capacities and "represented" by a cognitive system that exists independent of the world. Rather, the cognitive system "enacts" the world in cognition. But of course this does not mean that cognitive system constitutes the world in a vacuum. As Varela emphasizes, cognition depends upon the kinds of experience that come from various sensorimotor capacities of body, and these individual sensorimotor capacities are themselves embedded in a more encompassing biological context, i.e., they are formed through a history of structural coupling with environment (interference between the two autopoietic systems, cognitive system and environment).

For Varela consciousness has double sense: it encompasses both the psychological consciousness (environmentally conditioned information-processing system) and the lived, transcendental consciousness (self-referential autopoietic system). These two sides of consciousness are obviously not opposed. Instead, we continuously circulate back and forth between them. Varela led the way, by making this double sense of consciousness the focus of attention, to explore the possibility for circulation between cognitive science and phenomenology. But unfortunately, his attempt was interrupted by his death in 2001. Our task is to continue a program of research founded by Francisco Varela.

---

[38] H. Maturana and F. Varela, *Autopoesis and Cognition: The Realization of the Living* (Dordrecht: D. Reidel, 1980), p.51.

Chapter 11

The Three Species of Relevancy in Gurwitsch

Lester Embree
Florida Atlantic University

Abstract

Aron Gurwitsch's central category of relevancy is difficult for some to grasp. It is approached here through comments on its generic signification but chiefly through seeking its species and their examples. On this basis, not only can relevancy in general as well as its species be more clearly grasped via generalizing universalization, but the possibility of further phenomenology can be seen.

Introduction

Aron Gurwitsch once told the present writer that he considered his analysis of relevancy his greatest contribution. This analysis is central to his doctoral dissertation of 1929, "Phänomenologie der Thematik und des reinen Ich. Studien über Beziehungen von Gestalttheorie und Phänomenologie";[1] it is chiefly developed and extended in his masterpiece of 1964, *The Field of Consciousness*;[2] and there are pertinent passages not only elsewhere in *Studies in Phenomenology and Psychology,* but also in *Esquisse de la phénoménologie constitutive,*[3] *Human Encounters in the Social World,*[4] *Marginal Consciousness,*[5] *Phenomenology and the Theory of Science,*[6] and in still uncollected publications.

Yet Gurwitsch's account can be difficult to comprehend and therefore difficult to accept. This may be in part because its expression is dispersed, but it can also be due to the universality in terms of which it is often expressed. In the latter respect, a reader can crave the clarity that examples as well as the distinguishing of species can bring. There clearly are three species but it will be nec-

[1] Trans. Fred Kersten as "Phenomenology of Thematics and of the Pure Ego: Studies of the Relation between Gestalt Theory and Phenomenology," in Aron Gurwitsch, *Studies in Phenomenology and Psychology* (Evanston, IL.: Northwestern University Press, 1966), 175-286. Hereafter this collection will be cited as "SPP."
[2] (Pittsburgh: Duquesne University Press, 1964). Hereafter this text will be cited as "FC."
[3] Ed. José Huertas-Jourda (Paris: Librarie Philosophique J. Vrin, 2002). Hereafter cited as "EPC."
[4] Ed. Alexandre Métraux, trans. Fred Kersten (Pittsburgh: Duquesne University Press, 1979). Hereafter: "HE."
[5] Ed. Lester Embree (Athens, OH: Ohio University Press, 1985). Hereafter: "MC."
[6] Ed. Lester Embree (Evanston, IL: Northwestern University Press, 1974). Hereafter: "PTS."

L. Embree (ed.), Gurwitsch's Relevancy for Cognitive Science, 205-219.

essary to name them. The species and instances will be studied after some comments on a general expression.

I. Relevancy in General

Relevancy is something that occurs within the field of consciousness. Gurwitsch acknowledges that the notion of field comes from Gestalt theory, which received it from 20th century physics (SPP, 62), and also from William James, who held that the total field is structured into a focus and a margin. Gurwitsch, however, not only writes of "theme" instead of "focus," but also divides James's "margin" such that part of it is the "thematic field" and the rest is the "margin." Items in the thematic field "have to do with" or are relevant to the theme, while the marginal items are devoid of relevancy. His friend, Alfred Schutz used "relevance" and urged Gurwitsch to use "pertinence" in English, but Gurwitsch uses "relevancy" almost always ("pertinence" does occur, e.g., FC, 4); "*Relevanz*" is used in German).[7]

The following passage employs universal terms in describing what can be observed reflectively and, more specifically, in noematic reflection, which is to say reflection on things-as-they-are-encountered.

> The structure is realized at every moment of conscious life, whatever object (in the broadest sense of the term) presents itself through a given act. As every formal invariant assumes a specific form in every concrete variety in which it is embodied, so is relevancy specified in every case in which a theme appears in a thematic field. As actually experienced, relevancy is always qualified and specified in accordance with, and in dependence upon, the material contents of both the theme and the thematic field. (FC, 379)[8]

[7] Alfred Schutz and Aron Gurwitsch, *Philosophers in Exile: The Correspondence of Alfred Schutz and Aron Gurwitsch, 1939-1959*, ed. Richard Grathoff trans. J. Claude Evans (Bloomington, IN: Indiana University Press, 1989), pp. 151 and 350, cf. FC, 342. See also Lester Embree, "Everyday Social Relevancy in Gurwitsch and Schutz," *Annals of Phenomenological Sociology* 2 (1977), 45-61, and as "Theorien sozialer Relevanz: Aron Gurwitsch und Alfred Schutz," trans. Alexandre Métraux, in *Alfred Schutz und die Idee des Alltags in den Sozialwissenschaften*, ed. Walter M. Sprondel and Richard Grathoff (Stuttgart: Enke Verlag, 1979), 65-77.

[8] About "object" in the broadest signification, there is this early passage: "Object, here meant in an all-inclusive sense, comprises not only material things perceived, as well as remembered or imagined, but also mathematical relations, musical compositions, and theoretical implications" (FC, 4).

Some may have a problem here with the concept of relevancy being subsumed under that of "formal invariant" and being itself specified because there is no subsumption and specification in formal ontology. Yet the form in question would seem not to be the product of formalization, which abstracts from material contents, but more like shape as abstractly distinguished from color.[9] It might also be that, even though he is referring to material *eidē*, Gurwitsch is using "form" in a vaguely Platonic signification that includes colors and character traits as well as numbers. And if "form" is still bothersome, the reader might try substituting "structure," which the passage above uses synonymously.

Systematically, the question of what relevancy in general is precedes that of its species, but the species can be easier to grasp first. On both levels, what are called "material contents" and, in relation to them, "material relationships" are not only not abstracted from but are crucial. If one begins with a thing-as-encountered and abstracts from how it is given and also how it is posited, the "noematic nucleus," "sense," or, best, the "core" of the noema remains.[10] Gurwitsch writes that the "'materially determinate [*Sachhaltig*]' [is] understood in contrast to that which belongs to the characters.... In this sense, not only are terms such as 'colored,' 'rough,' etc., meant under the heading of the materially determinate, but also 'property,' 'figure,' 'relation,' and so forth" (SPP, 185 n. 21). Phenomenological description can be done in formal-ontological and also in material-ontological term and material contents and relationships call for the latter.

Gurwitsch creatively employs concepts from Gestalt theory to go beyond Husserl in describing the noematic core. Thus there are "material relations *(sachliche Bezüge)*" (SPP, 199), also expressed as "material relationships" (e.g., FC, 351),[11] and he approvingly reports that "Köhler defines—though only in passing—the material relation between two things[12] as a '*context owing to the properties* of those things themselves ..., not their frequent occurring together,

---

[9] "According to Husserl, formalization consists in replacing terms which have material content by 'empty' ones, that is to say which are defined exclusively by their function for a certain operation, in the case under discussion the activity of going from item to item. This analysis provides a phenomenological account of formalization and also, closely related thereto, of the phenomenological origin of the notion, so important and fundamental for Husserl as well as Gonseth, of 'something at large' ('*Etwas überhaupt*,' '*quelque chose en général*')." Aron Gurwitsch, ed. Lester Embree, "On Thematization," *Research in Phenomenology*, 4 (1974), p. 46.

[10] This approach is presented in some detail in the introduction to the present volume.

[11] Gurwitsch revised the German translation of FC before he died. This is rendered there as "*die Beziehungen zwischen den beteiligten Sachgehalten*" (Aron Gurwitsch, *Das Bewusstseinsfeld*, trans. Werner D. Frölich [Berlin: Walter de Gruyter, 1975], 283).

[12] *Gurwitsch's note*: "The reference is not to thing in the proper sense but in the frequent usage where it signifies object, moment in the object, property of the object, and the like."

successively or simultaneously.' By this, however, he designates what is characteristic of and essential to the Gestalt connection as such in opposition to the mere 'sum'" (SPP, 212).[13] Then again, there is a "Gestalt connection between theme and thematic field" (SPP, 210), although it is also true that "unity by Gestalt-coherence underlies and renders possible unity by relevancy" (FC, 358, emphasis altered).[14]

Relevancy is the genus of material relationship that is most referred to, but there are others. The chapter on "Types of Gestalt-contextures" can be taken as about material relations. Thus, one can see two points on a homogeneous background as unrelatedly side by side or, if they are fairly close together, one can see them as a pair.

> If the points are seen as forming a dyad, the interval between them does not appear as a gap, but as their connecting link. However, it is not as though the points were originally disjoined and had to be brought into connection subsequently. A connecting link denotes an essential constituent of the appearing structure: an interval delimited and encompassed by, there-

[13] "The thematic field itself is a context of 'objects' intrinsically related as materially belonging together. The form in which all items belonging to the thematic field are connected with both each other and the theme is not the form of the 'and-connection' in which anything whatever can be connected with anything else whatever, however disparate and for which material relationships or relations of relevance are of no account. Since the items in question belong together on the basis of their properties and their *material* contents, there obtains between them a 'Gestalt connection' (*Gestaltverbindung*) which in every given case is specified in accordance with the specification of *those material relations.*" (SPP, 203)

[14] In different Gestaltist terms, Gurwitsch also refers to three types of conjunction between items within the field of consciousness. One might suspect that all of them are material relations too, but this is not exactly the case: "The *first type of conjunctions*, ... is that *between the constituents of the theme*. For conjunctions of this type we have proposed the term *gestalt-coherence*. In the *second place*, there are the *conjunctions between the theme and the thematic field* and also *between items of the thematic field*. Based upon the material relationship between the contents involved, the conjunctions of this type constitute *unity of context* or *unity by relevancy*. ... *Finally, conjunctions between both the theme and the thematic field*, on the one hand, and on the other, *the margin, are of mere temporal nature*. They consist in either the simultaneous or immediately successive occurrence of the respective acts" (FC, 352-353). Thus, at least one type of conjunction is not founded on a material relationship, but at least one type is. What all the variously conjoined items have in common is co-presence, which is not noematic but noetic: "The co-presence of marginal data with the theme and the thematic field is only a relation between the mental acts, i.e., the experiences having such matters as intentional correlates, and this relation consists solely in these acts happening to occur together" (MC, xliv; cf. FC, 344).

> fore, conjoining, a pair of terminal points. Perceived as delimited and enclosed, the interval appears set off from the field with a perceptual emphasis regarding the rest of the field. The latter forms the background from which this structure emerges. (FC, 107)

Here the material relation is spatial. Perceived spatial relations can be more or less close. Temporal intervals also occur (FC, 108). And then there is the figure/ground structure, which cannot only be seen in space but also heard in time (FC, 113) and which Gurwitsch asserts is a special case of the theme/thematic field relationship, i.e., relevancy (SPP, 204 n. 49; SPP, 211; FC, 356-57). In other words, material relations of the figure/ground variety are of a sort of relevancy, while relevancy itself seems a kind of material relationship on a par with Gestalt contexts of the spatial and temporal sorts.

II. Naturalistic Relevancy

If the figure/ground relation is a species of relevancy, what examples of this relation are offered? The perspective of the analysis remains one of noematic reflection, i.e., reflection on the thing-as-encountered. In this perspective, Gurwitsch contends that Edward Rubin's analyses of plain drawings extend to material things, e.g., houses among other houses, cars in the street, etc. (FC, 320), which are of course functional or cultural objects, something considered in Section III, but he clearly discusses such things with their functional or cultural characteristics abstracted from and the manner of givenness kept constant as visual perceptivity and thus focuses on their core determinations, e.g.,

> The independence of the figure with respect to the ground appears in the role of the bounding contour which demarcates and sets off the figure from the ground so as to make it emerge as a self-contained unit. That independence also appears in the thing-character of the figure as contrasted with the ground which presents the characteristics of "stuff" or "material." … Every such figure appears as displaceable on the ground from which it actually emerges and also as transferable from one ground to another (in the case of a black figure from a white ground to a red one), without any impairment of its phenomenal identity. (FC, 357; cf. FC, 110 ff. and SPP, 208)

There can be many other examples, including those pertaining to sense modalities other than vision. How the figure/ground structure relates to other species of relevancy will be considered below. Here it may still be remarked that in the naturalistic attitude requisite for the naturalistic sciences, which is to say when cultural characteristics are abstracted from, this is the only species of relevancy that there is. Gestalt theory seems chiefly conducted in the naturalistic theoretical attitude, but this is not the case for Gurwitsch.

III. Practical Relevancy

There are two other species of relevancy for Gurwitsch. He often refers to the scientist walking in the street thematizing a scientific question and having a thematic field composed of propositions relevant to it. That species will be returned to in Section IV. Here the concern is with relevancy in everyday practical life where the things that are encountered retain their uses or functional characters and can thus make up the subject matter of the cultural sciences (cf. PTS, 177-78).

In his classes, Gurwitsch often urged students to study an author's examples. Most of his own examples of relevancy are not only of an everyday sort, but also relate to writing, the utensils involved, and the situation in which it is done. In the 1950s he wrote of the "sphere of action to which … a typewriter belongs" (FC, 100), while earlier there were inkwells, desks, paper, and pens. The following is a statement about indoor practical situations, including those in which one might write, i.e., a study.

> Sitting in a room, we are aware of the things behind our back through horizonal consciousness as things which fit into a room of this kind, whether or not these things present themselves with full determinacy regarding their particular details. In this case, the experience of context rests upon the experienced suitability of the things in question to the purpose of the room, the awareness of this purpose being conveyed by those things which are given in actual perception. (MC, 47)

Upon reflection, situations of which there is consciousness can be recognized to have theme/thematic field/margin structures. "Theme" does not have the musical signification, but "is meant to denote that with which the experiencing subject is dealing, that upon which he is dwelling, that to which his attention is drawn in a given act of experience" (FC, 323 n. 33, cf. SPP, 177). Some may be inclined say "object" (or "thing") rather than "theme," but there are also relevant objects and marginal objects.

> More than a theme is always encountered: I observe, e.g., the inkwell on my desk. Round about it lie pencils, paper, books, etc.; they are co-given while I am directed to the inkwell in the mode of cogito: they form the "objective background" of the inkwell. In this "objective background" there appear diverse things, among them those just named; but even more: there appears to me through the window part of the front of my neighbor's house. (SPP, 196)

Already it is clear that more than merely naturalistic figure-ground relations are involved. Beyond the thematic field is the margin:

> Being in a practical situation and making use of the functional objects involved in it can become an absorbing activity, but never to the point where the existence of the world surrounding the current state of affairs is completely forgotten or lost from view. When writing, for instance, or making something, one retains a certain awareness of the surroundings, of being in this room of this house which is on this street of this city, with something going on outside, but without a knowledge of exactly what or, at least, without any feeling of a need to know. (PTS, 173 = EPC, 323)

Noetically speaking, i.e., with respect to the encountering rather than the thing-as-encountered, it is already clear that there is practical activity as well as utensils involved in such a practical situation, e.g., the dipping of the pen in the ink. *Pace* Descartes as well as Husserl, action, i.e., a volitional thematic consciousness, is a type of "cogito." Then there is how action involves useful or functional objects.

> As a writing-utensil, the pen refers to writing, i.e., to that for which it serves in a manner befitting it. Such manipulating is, however, never a mere dealing with something. In writing, I do not take pen in hand in order to manipulate it as a specific utensil. If I do that, then I do not use it as the utensil which it is. In meaningful use, I use the pen in order to write with ink on paper. Similarly, the pen refers to paper, ink, inkwell, table, etc. It "belongs" to these requisites of writing, which, for their part, "belong together" with respect to writing and form a "utensil totality" with a "utensil structure." The *second kind* of utensil-reference concerns this utensil-totality; the particular utensil is related to this structure in the sense that it becomes

> the utensil that it precisely is with respect to the structure. (HE, 70; cf. HE, 67; FC, 231)

Correlative to writing, "my pen is something for writing, and not a black thing of such and such length and almost cylindrical shape."[15] The pen is thus the utensil prior to any abstraction used to access the material contents of the noematic core and thus consideration of it as a naturalistic thing. The instrumentality of an instrument or utensil can change (FC, 49) and one can think of its usefulness in another situation or remember the role it once played in one's life (SPP, 202). Gurwitsch also refers to using a hammer, and asserts in general that "the object is not perceived as a mere thing, but rather as an object of use to be manipulated in a determinate manner" (FC, 231; cf. PTS, 171; HE, 77).

If one were to perform the abstraction from the instrumentality or use, the hammer becomes "a thing of a certain length, breadth, thickness, and weight, of such a form, with a particular color, and so on" (PTS, 171; cf. HE, 43), i.e., one can discern "material properties of the physical thing … on the basis of which, in a certain situation, in this or that way, the physical thing can be used" (HE, 44). But Gurwitsch also refers to Husserl's description of the natural attitude in §27 of *Ideen I* in order to highlight how things are always already fitted with values and uses (*idem*); accordingly, "if the hammer is perceived as a tool, with the sense of its instrumentality, it is thus perceived previously to, and independently of the process of analytical explicitation" (FC, 235).

Here it may be remarked that for Gurwitsch and most phenomenologists, the sociocultural and practical lifeworld has at least a methodological priority over the universes constituted in the sciences. This is because this world is where one must begin from even if many forget about it.

Returning to the writer's study, the inkwell looks different when it is put on the piano instead of the desk where it belongs (SPP, 206), but it certainly retains its strict identity as the same inkwell (SPP, 207). (Although Gurwitsch does not mention it, an inkwell can belong on the piano for the composer or arranger of music.) In general, a theme can be moved from one thematic field to another (SPP, 208). Moreover, "thematic fields may so differ from one another as hardly to have in common any material element of content, except the theme. To illustrate: we may think of Goethe one time as a poet and creative genius of the German language, another time as a minister in the service of the Duke of Weimar" (FC, 322).

The thematic field can be broadened as new items appear (SPP, 223). One can withdraw from one theme and take up another item as theme in the same thematic field (FC, 371) and something in the margin can become a theme. Overall, the world is a set of potential as well as actual themes (HE, 41). Finally,

---

[15] "*Mon stylo est quelque chose pour écrire, et non pas un corps noir de telle et telle longueur et de forme presque cylindrique*" (EPC, 324).

and referring further to manners of givenness other than the perceptual, the structure whereby themes always have thematic fields also holds good in remembering, expecting, and phantasy (SPP, 200). The general points that have been mentioned can easily be related to the example of the study and writing whether with a pen or a typewriter. So a book may be physically on the desk but marginal during writing, but when the writer wants to find a passage, the book enters the thematic field as a possible source.

Before going further, it may be asked if there is specifically social relevancy. Gurwitsch certainly mentions animals and fellow-men[16] and there are other humans in what he calls "spheres of life," which are sub-orders of the paramount reality that is the "perceptual world" as a whole (FC, 383). These include the spheres of family life, politics, and professional activity and these are unified by "specific relevancy-principles":

> As to the spheres of professional life, the relevancy-principles involved vary obviously according to the nature of the professional activity. Whatever its nature, every sphere of professional life can be described as a system of purposes, plans, projects, designs, ends, means, and actions. All these refer to one another, each has a certain determinate place within a hierarchical order and organization. Fellow-men encountered within the sphere of professional activity appear in, and appear as defined by, their roles within that sphere. They present themselves as students, teachers, doctors, patients, employers, employees, business associates, customers, officials in whom authority is vested, and the like. Similarly, things are also experienced under the perspective of situations of concrete action in which they serve as instruments and tools, thus appearing with reference to purposes relevant to those situations. (FC, 383)

As shown, the basis for the close involvement of humans in social roles with utensils can be found in *Human Encounters in the Social World*, but Gurwitsch unfortunately does not analyze the social in its own right as extensively as he might have. Nevertheless, it seems clear that others are objects that can be thematic, relevant, and marginal in fields of social consciousness.

---

[16] "At every moment of conscious life, we find ourselves within the perceptual world, amidst things and objects of greatest diversity: natural objects as well as objects of value or cultural objects, inanimate things as well as animals, fellow-men to whom we are standing in various relationships" (FC, 161, cf. FC, 221).

How is there relevancy in fields of functional objects? This seems best called "practical relevancy." Fairly clearly, relevancy is here a matter of teleology with actions as well as implements relating to a theme in means-end relations. *To be useful for a thematic purpose is to be relevant.* Here Gurwitsch's reflective analysis converges with the use of "relevance" in current ordinary American English. Reflecting on what we are doing, we can see that if our pen runs out of ink, our thematized goal is a freshly dipped pen, and the inkwell in which to dip it is then relevant. Relevancy is defined as "a relationship founded upon the material contents of the items concerned, which have something to do with one another by virtue of their material contents" (FC, 394). This would cover ink as a fluid that can dry, the container for it, its adhesion to pen points, and something to which it can be applied and be absorbed. But there are limitations as well that are due to material contents and relationships, for one cannot hammer nails with a banana or write well on flowing water.

Finally, "for a datum to belong to the thematic field [of] a given theme, it must satisfy the condition of relevancy to the theme, a condition concerning the material contents of both the theme and the datum in question" (FC, 414). Besides material contents, material relations of the sorts described in Section III must be such conditions, i.e., reflectively observable spatial, temporal, and figure/ground relations within the noematic core. These are foundational for the uses or functions of the functional objects.

In many cases, however, causation should be recognized as underlying means-end relationships. Gurwitsch actually says little about causation.[17] There is, however, this passage in a practically forgotten review of Hans Kelsen's *Society and Nature* (1943).

> Lévy-Bruhl's "participation," the principle of causality understood to purport an internal connection between two events which succeed each other in time, and, finally, the idea of functional dependency which, as Kelsen points out, is the form in which "causality" is conceived in modern science, are specifications of the quite general concept that there is context and contexture in universe, that things hang together. This general concept of context at large is a necessary and indispensable condition of the experience and existence of a possible universe, however restricted and confined a particular "universe" might be.[18]

[17] See the Introduction to this volume, Section I, for the model of explanation that Gurwitsch accepted from Gestalt theory.
[18] *Isis*, Vol.. 36 (1946), 145.

Most of Gurwitsch's statements about relevancy are rather general, but in a practical case, one can ask what the theme is; what is relevant to it; and what material contents and relations in the noematic cores are conditions for the uses reflectively discernable in the lifeworldly objects from which all inquiry begins. Furthermore, the consideration of the purposes of situations, institutionalized structures such as rooms (studies, kitchens, etc.), and spheres of life help further specify and exemplify this practical species of relevancy. In these respects one is implicitly invited to recognize and describe subspecies of this species.[19]

Relevancies of this second species occur in fields of consciousness composed of functional objects. Here it is especially interesting how such things (humans in spheres of life included) acquire functional characters or, if they always already have them, then how these noematic characters change. After all, there once was a time in every subject's life when pens and inkwells were unfamiliar, but then, probably through observation and imitation, the pertinent skills and instrumentalities were established.

> Functional characters accrue to objects in situations of concrete action in which the subject manipulates the object, learns to handle it in a determinate manner, to use it for a certain purpose in connection with other objects, and thus acquires a certain mode of action. Such acquisition will henceforth co-determine future perceptions. Suppose the same or similar object is reencountered after the acquisition of the mode of action. The later perception occurs under different conditions, the difference consisting in that the mode of action has been aquired. Such acquisition modifies in a certain manner the psycho-physical system, so that processes occurring after that event differ from those which occurred before. Accordingly, through the latter perception the object does not appear as it did through the earlier one. (FC, 98-99)

It is not stated by Gurwitsch, but still seems plausible that just what is relevant to what in typical situations of action has the same explanation. Thus theme-thematic fields combinations would also arise and change. Indeed, many matters of culture, including above all utensils and social roles, but also including variations due to social class, historical period, gender, age, pathological condition, etc., can be approached through experimentation as well as through empirically informed reflective analysis. And a reason why Gurwitsch's category can seem overly abstract is that the concrete variety that comes under this species is so vast.

---

[19] This can be said to be done by Louis Sass in his chapter in the present volume.

IV. Formal-Scientific Relevancy

Gurwitsch was of course interested in the sciences. Indeed, his interest in lifeworldly practical consciousness is arguably for the sake of science theory. He discusses history and biology as well as mathematics, physics, and psychology,[20] but the special science in relation to which he chiefly discusses the theme/thematic field structure is formal logic. Insofar as the non-formal sciences must be logical, his account of logic holds for them and, in that case, the third species of relevancy can be called "formal-scientific."

Gurwitsch's analysis of this species is concentrated in the chapter on "Context in Logic": "Following a chain of arguments, we consider one proposition after the other. Each proposition, in turn, becomes our theme" (FC, 325). Going beyond the Husserlian analysis of non-core components of the thing-as-encountered into manners of givenness and types of positionality, he writes, "We submit that the characters here in question, [i.e.,] derived from, leading to, compatible with, and contradictory to, must be recognized as noematic characters" and he calls them "contextual characters" (FC, 327). These can be abstracted from to reach the noematic cores of propositions. When that does not happen, "the thematic field consists of propositions" (FC, 329), which can be investigated through noematic reflection, although thinking is more often conducted in a straightforward attitude: "While formulating propositions or proceeding from one proposition to those following from it, our interest usually is directed towards the states of affairs to which the proposition or system of propositions refer rather than the propositions themselves, their logical structure and concatenation" (*idem*). Logical relevancy is also clear in the following passages.

> When a proposition is experienced as derived from, leading to, compatible or incompatible with, other propositions, such contextual characters are to be considered as specifications of the phenomenon of context or pertinence in general. *The phenomenon of context or pertinence in its unspecified or rather prespecified form is,* we submit, *a necessary condition of logical relationship in the proper sense.* (FC, 332)

Similarly,

> If the given proposition points and refers to a system of propositions, it is because the material terms of the former have something to do with, and are in some way or other relevant

[20] See Lester Embree, "Aron Gurwitsch's Theory of Cultural-Scientific Phenomenological Psychology," *Husserl Studies*, 19 (2003), 43-70.

> to, the material terms appearing in propositions belonging to the system. Herein consists the phenomenon of context. *Consciousness of context or of pointing references is one and the same as consciousness of relevancy*, however vague and indistinct. (FC 335).[21]

There is much more about logic in Gurwitsch, but this should suffice to show the third species of relevancy.[22]

V. Relevancy and Orders of Existence

Gurwitsch's masterpiece culminates in a philosophical analysis of "orders of existence." There is much of great interest there, but present purposes only require showing that, as one chapter heading has it, "relevancy-principles [are] constitutive of orders of existence" (FC, 391 ff.).[23] To exist, for him, is to belong to an order of existence. What is such an order and which ones does he mention? Two passages are relevant to these questions:

> To denote the widest range of items and data referred to through the consciousness of indefinite continuation of context, we shall use the term "order of existence." By the thematic field, we mean that part of the order of existence immediately adjoining the theme and whose components are experienced to be of direct concern to, and of immediate bearing upon, the theme. Of course, it is almost impossible to draw a sharp line of demarcation between the thematic field and what lies beyond the thematic field and still belongs to the order of existence in question. ... What we denote as order of existence is, in the final analysis, an indefinitely extended thematic field. (FC, 381)

> Orders of existence within the meaning of our definition are the "natural groupings" in which things present themselves in pre-scientific and pre-theoretical experience as well as the explanatory systems constructed in the several sciences for the

[21] The "material terms" mentioned here are not the "material contents" of the noematic core, but rather the terms abstracted from in formalization (FC, 331).

[22] Cf. Lester Embree, "Gurwitsch's Theory of Logic." *Social Research* 42 (1975): 129-37.

[23] Cf. Aron Gurwitsch, "The Problem of Existence in Constitutive Phenomenology" (1961), SPP 116 ff.

> sake of a rational explanation of the world, material, historical, and social. We must also mention purely ideal orders of existence, such as logical systems, the several geometric systems, the system of natural numbers, the generalized number-systems, and so on. Finally, there are the universes of artistic creation like the universe of music. (FC, 382)

Gurwitsch goes on to posit objective time as the relevancy principle for reality in general and the "perceptual world"[24] is asserted to be the fundamental or paramount stratum of reality. The sub-orders called "spheres of life" referred to above in the discussion of social relevancy are within this "perceptual world." Objective time is not so-called "phenomenal time," which is the inner time of the stream of consciousness, but the time in which grass grows and history goes on, and, "on the basis of the one, unique objective time, in which the life-histories of all persons take place, all the spatial surroundings of those life-histories are unified into one all-encompassing order of existence, namely the one real, objective, spatio-temporal world, the life-world of all human beings communicating with each other either directly or indirectly" (FC, 387, emphasis altered).

But there are also orders that are not part of this reality, i.e., those of imagination (FC, 388 ff.) and those composed of ideal objects, such as scientific theories made up of logically related propositions. One can thus posit an ultimate genus that can be called "objective being" and that includes non-real as well as real things. Moreover, there are not only spatio-temporal realities, but things within the real that are temporal but not inherently spatial, i.e., consciousness viewed from the transcendental point of view.[25] Under that genus, the specification and exemplification for logical objects and also for naturalistic objects in spatial, temporal, and figure/ground relations has been seen. What of the fundamental world of functional or cultural objects? Objective time alone is the relevancy principle for reality in general and thus includes consciousness that is temporal and intentional, but not spatial. Yet, "the real material world, the perceptual world, presents spatio-temporal, and not merely temporal, unity and continuity" (FC, 393).

What have been emphasized thus far are orders of existence as extended thematic fields that are actual, but there are also orders to be found in the margin, which are potential thematic fields. When the theme is a scientific theory or something in a world of imagination, such as a novel, all of the following orders

---

[24] The "perceptual world" is the cultural world composed of functional and other cultural objects in *The Field of Consciousness,* but in the later essays is composed solely of naturalistic objects and it is thus contrasted with the cultural world.

[25] Cf. Lester Embree, "Gurwitsch's Phenomenology of the Margin, Body, and Being," MC, xxxiv.

are marginal: "1. *The stream of our conscious life*; 2. *our embodied existence*; 3. *the perceptual world*" (FC, 415; cf. MC, *passim*), and when the theme is in one of these orders, the other two are marginal.

*
* *

In Gurwitsch's oeuvre, relevancy is actually specified and exemplified in three ways. For the noematic cores abstractly discernable in things-as-encountered, it is a matter of figure and ground. When the cultural or functional characteristics are not abstracted from, it is a matter of practical contexts including above all means and ends, which can be founded upon the spatial, temporal, and also causal as well as figure-ground relations reflectively observable in the cores of things-as-encountered. And for the relations of propositions in the sciences, they involve above all the contextual characters fundamental for how propositions can combine into arguments. Most of the time, Gurwitsch refers to what can be called "immediate" relevancy, but he is fully cognizant of the indefinitely extended contexts in which real and ideal orders of existence and indeed being in general are constituted.

Chapter 12

Kinds of Knowledge: Phenomenology and the Sciences

Osborne P. Wiggins, Jr.
University of Louisville

Michael Alan Schwartz
Case Western Reserve University

Abstract

For Aron Gurwitsch, phenomenology was a science, albeit of a special kind. It was neither an empirical science, like physics or psychology, nor a formal science, like mathematics. Our essay seeks to clarify some features of this special science and its relationships to other kinds of knowledge. The relationship of phenomenology to empirical psychology is explored by following Gurwitsch's amplifications of the rejection of the "constancy hypothesis" by Gestalt psychologists. The main components of a distinctively phenomenological method are then explicated. In Gurwitsch's later articles, under the influence of Husserl's *The Crisis of European Sciences and Transcendental Phenomenology,* there emerges a phenomenological theory of all the sciences. We sketch that theory, focusing especially on Modern "naturalism." This focus allows us to draw implications for the contemporary project of "naturalizing the mind."

Introduction

Aron Gurwitsch always sought to understand the relations of the sciences to one another (1964, 1966, 1974). In particular he sought to understand the relation of phenomenology to psychology since they both purported to study the same topic and since Gurwitsch thought they could learn from one another (1964, 1966). He was convinced that phenomenology could learn especially from Gestalt psychology. He perceived in Husserl's phenomenology certain traditional theoretical prejudices that he believed Gestalt psychology had at least partially overcome. Hence he tried to forge links between Gestalt psychology and Husserlian phenomenology. In this essay we shall address only one aspect of this attempt: Gurwitsch's phenomenological interpretation of the Gestaltist criticism and abandonment of the "constancy hypothesis."

L. Embree (ed.), Gurwitsch's Relevancy for Cognitive Science, 221-233.

In Husserl's *The Crisis of European Sciences and Transcendental Phenomenology* (1962, 1970) Gurwitsch found the key to an area that he himself had sought to explore: a phenomenology of the intellectual genesis of the sciences. This genesis was partly historical, but it was also partly eidetic. Gurwitsch then borrowed Husserl's key and further systematized this phenomenology of the relations of the sciences to one another. We shall discuss this topic in the second half of our essay.

This discussion will allow us to pinpoint those limited areas in which "naturalism" can claim legitimacy.

I. The Constancy Hypothesis and its Generalized Formulation

Aron Gurwitsch interpreted the abandonment of the constancy hypothesis by the Gestalt psychologists as an incipient phenomenological reduction. The constancy hypothesis was a theoretical preconception pervasive in psychology around the beginning of the twentieth century. Most briefly stated, the constancy hypothesis entails that there exists *a one-to-one correspondence* between physico-chemical stimulations of the sensitive surfaces of the human organism and neural events in the brain and that there also exists *a one-to-one correspondence* between these neural events in the brain and sensations or sense data occurring in human experience. Because of these two correspondences, there exists a one-to-one correspondence between stimulations of the sensitive surfaces of the body and sensations in the mind.

The constancy hypothesis was, however, a particular instance of a more general theoretical preconception. From what he writes one can tell that Gurwitsch was aware of this more general preconception although, as far we know, he never explicitly formulated it.

We shall try to formulate it here: *one is justified in deriving inferences about human experience from theoretical views about the human brain.* The telling term here is "justified." It is a question of what *justifies* us in postulating the actual occurrence of specifiable mental processes. The constancy hypothesis maintains that theoretical views about brain events sufficiently justify the postulation of particular psychological events.

II. A Consequence of Accepting the Constancy Hypothesis

For the psychologists who accepted the constancy hypothesis, one of the results of such acceptance was a *discrepancy* between the experiences their experimental subjects actually had and the experiences that, according to the constancy hypothesis, they "should" have had. One way to resolve this discrepancy would, of course, have been to abandon the constancy hypothesis and to accord primacy of place to the subjects' experiences. The psychologists who accepted the constancy hypothesis refused, however, to relinquish it despite the clear discrepancy. What these psychologists did instead was insist that the subjects were indeed having the experiences entailed by the constancy hypothesis but that these experiences were simply "unnoticed" ones. The subjects were having experiences, so it was claimed, of which they remained unaware. Saving the constancy hypothesis, therefore, required the postulation of an entire realm of "unconscious mental processes."

It would also seem that what we have called "the general theoretical preconception" of which the constancy hypothesis is a special instance would regularly require the postulation of " unconscious mental events." If one is justified in deriving inferences about human experience from theoretical views about the human brain, then, when the neuroscientist's views imply it, one can postulate the existence of mental processes even though the subjects are unaware of having them. And indeed today there are a number of cognitive scientists and psychologists who are drawn by their neurological views to postulate the occurrence of unconscious mental processes. Once, of course, such postulates are admitted, there can be no limit to their variety; and consequently, one cannot restrict the variety of theories of mind that are generated.

III. One Consequence of Relinquishing the Constancy Hypothesis and its Generalization: Mental Life as an Autonomous Field of Study

Let us now return to the "generalized form" of the constancy hypothesis. That generalized form, it will be recalled, stated that one is justified in deriving inferences about human experience from theoretical views about the human brain. The relinquishment of the constancy hypothesis, then, consists in *denying* that one is justified in making inferences about experience from views about the brain. The implication that Gurwitsch draws from this denial is that human experience must be investigated *in its own right,* independently of any claims or preconceptions from other areas or theories.

The question then arises of the method best suited to investigate this domain of human experience. Gurwitsch thought that this method was the phenomenological one developed by Edmund Husserl. Husserl's phenomenological method has six main components. We shall here sketch five of them.

A) Reflective. The phenomenologist proceeds by taking the processes of his or her mental life as examples of the kinds of mental processes he or she wishes to study. The phenomenologist thus reflects upon the processes occurring in his or her own mental life and examines them and their intended objects.

B) Intuitive. The phenomenologist examines only those mental processes that can be *directly observed,* only those mental processes that are *directly given* to his or her reflective gaze.

C) Descriptive. All claims to knowledge made by the phenomenologist are *descriptive* claims, i.e., claims that express only those features of mental processes and their intended objects that are directly observed (i.e., intuited) in reflection.

D) Eidetic Reduction. The phenomenologist seeks to define *the essential features* of different kinds of mental processes. The essential features of a class are those features without which something could not possibly be a member of that class. In other words, the essential features of a class are the features that are *necessary* for class membership. Necessary features differ from factual features. Factual features are those features that a class of things has *as a matter of actual fact.* Essential features are features that a class has *of necessity.* To define the essential or eidetic features of a class of mental processes is to define the range of possible members of that class. Hence the phenomenologist, in order to define the essential features of a designated kind of mental process, must examine the range of possible exemplars of that class.

The phenomenologist must therefore move beyond the range of the actual in examining the sort of mental process that he or she has elected to study. He or she must move into the realm of the possible exemplars of that class. This move can be effected by systematically disregarding any distinction between possible and actual exemplars of the general kind of mental process under study and by being sure to include in one's reflection on exemplary mental processes purely possible ones. Such a methodological move is called "eidetic reduction." Hence when one reflects directly upon one's own mental life, one views the mental processes he or she finds there as purely *possible* mental processes. That is to say, he or she sets aside any presumption that would lead him or her to distinguish between actual and possible examples of the general class of mental process under study. Each mental process he or she reflects upon is now viewed as a *possible* one, no more significant for the purposes of study than any other possible example of the same class.

E) Free Variation in Imagination. The phenomenologist begins to systematically modify and vary the features of the exemplary mental process, now considered as a purely possible exemplar of the class under study. As these features are varied, the phenomenologist always keeps in view the class that it exemplifies. At some point in this variation procedure, the mental process will be directly seen to lose its membership in the chosen class. This directly observed loss of membership indicates that the feature that was varied was necessary for membership in the class. If loss of the feature entails loss of class membership, then the feature lost was essential for class membership. A feature without which the mental process would not be a member of the class is a feature that is essential for membership in that class. The phenomenologist thus has direct evidence for believing that the feature varied is essential to the general class of mental processes under examination. He or she can therefore tentatively claim that he or she has found an eidetic feature of that general kind of mental process.

Notice that the method implies that any claim regarding a feature of mental life cannot be accepted if that feature cannot be directly observed in reflection upon one's own mental life or its variants. Toward any claim that cannot be verified through such direct reflection the phenomenologist must remain strictly neutral. This is so because all phenomenological assertions must be *descriptive* ones, i.e., assertions verified through direct reflection upon one's own mental life or its imagined variants. On the other hand, the phenomenologist must accept (at least tentatively) as true any claim that can be verified through such direct reflection. With this method we have thus secured for investigation *in its own right* the domain of mental life.

We may now pose the following question: What can the phenomenologist's stance be toward claims about mental life that emerge from other disciplines, for instance, cognitive science or psychology? Gurwitsch has answered this question explicitly with regard to Gestalt psychological theory. He asserted that phenomenology could appropriate propositions from Gestalt theory only if these propositions can, through direct reflection, be confirmed as *descriptive* claims about mental life. In other words, statements from other disciplines can be phenomenologically accepted only if they can be verified through phenomenological procedures. Gurwitsch thought that the Gestalt concepts he called "Gestalt-reference" and "Gestalt-contexture," as well as the Gestaltist distinction between figure and ground ("theme" and "thematic field"), could be phenomenologically verified. He believed in fact that they denoted universal and eidetic features of mental life.

Notice, however, that this method requires that all phenomenological assertions be *eidetic*. Empirical disciplines, such as cognitive science, can investigate the features that mental life has *as a matter of actual fact.* Such features may also be *general* ones. The *methods* for studying such empirical features would simply not be phenomenological methods. In that case psychology and cognitive science would stand to phenomenology as "sciences of fact" stand to "sciences of essence": phenomenology would study the larger, more inclusive realm, the realm of the possible, while psychology and cognitive science would study the more limited realm, the realm of the actual.

IV. Naturalism and the Sciences of Mental Life

With the rise in Europe of the Modern mathematical science of nature in the seventeenth century, there emerges what Gurwitsch calls "a cleavage" between the world as it presents itself to us in our sensory experience and the universe as portrayed by this mathematical natural science (Gurwitsch, 1974, 34-35). With the emergence of this cleavage between the sense-perceived world and the scientifically conceived universe, a historically momentous ontological view develops. This ontological view Gurwitsch calls "naturalism." Naturalism is the following position:

> The universe is not as it appears in common experience and its nature and structure do not lay themselves open to perception; on the contrary, they must be uncovered by means of mathematical notions. In reality, then, the universe is not as it seems to be but as it is conceived and constructed by the mathematical physicist (Gurwitsch, 1974, 36).

Naturalism thus solves the problem posed by the "cleavage" between the sense perceived world and the scientifically conceived universe by attributing *objective ontological reality* solely to the latter. However, some ontological locus must be assigned to the sense perceived world since it stubbornly continues to exist, at least in our everyday experience. Beginning with Descartes, the thinkers endorsing naturalism succeed in finding a place for it: the entire sense perceived world, insofar as it differs from the scientific universe, is a mere *subjective appearance,* a more or less coordinated set of "ideas," "sensations," "sense data," or "mental representations." In any event, it is a "reality" that exists only in the mind (Gurwitsch, 1974, 210-240).

By the nineteenth century naturalism had more or less triumphed in Western cultures. Among its many offspring is a particular conception of the hierarchy of the sciences:

> Mathematical physics is the fundamental science of reality.
>
> Based on mathematical physics is chemistry.
>
> Based on chemistry and physics is biology.
>
> Based on all of these natural sciences are psychology and the other human sciences (Gurwitsch, 1974, 133-137).

Gurwitsch notes that this hierarchy carries the ready potential to lead theorists to *reduce* the concepts and laws of any science other than physics to the concepts and laws of the next most fundamental science. It also entails that a scientist working at any level must assume as given the accounts of reality found in the more fundamental sciences (Gurwitsch, 1974, 132-149).

For psychology, the science of the mind, this hierarchy has decisive consequences. Modern psychology presupposes naturalism as the basis of scientific intelligibility (Husserl, 1970; 20-100; Gurwitsch, 1966, 96-101). The influence of naturalism can be demonstrated by noting the form that psychology assumes an adequate explanation of human experience must take. From the outset the psychologist is seeking a theoretical account that presupposes the truth and explanatory power of naturalism. Stated in other terms, naturalism provides the principle of adequacy of a psychological account of human experience. That account must take the following form. Events in the human mind must be explained by correlating them with non-mental events in the human nervous system. Events in the latter are thought to be explained by correlating them with both other events in the nervous system itself and with still other events in the human organism. Events in the nervous system and organism are assumed to be characterizable in the terms supplied by neurobiology, physiology, molecular biology, and related natural sciences. Events in the human organism are to be accounted for by referring to the structure of the organism itself as it is causally affected by events in the organism's environment. Events in the organism's environment are assumed to be explainable by physics, chemistry, and other natural sciences. All of these events form a unified system in the sense that they are to be interpreted by the laws and concepts of natural science as interrelated causal processes.

Hence, as Gurwitsch has written,

> Modern psychology has developed not only in historical continuity but also in *logical continuity* with modern physics. What must be stressed is...the reference to physics in the very formulation of psychological problems, especially problems concerning perception. To account for perception, the psychologist accepts, and starts from, the universe as conceived in physical science and considers the human organism as a physical system acted upon by physical events. Independently of any theories to be advanced, the very problems that the theories of psychology are meant to solve are determined by allowance for the science of physics (Gurwitsch, 1966, 99).

V. A Phenomenology of the Sciences

Naturalism, however, is not the only view that can make sense of "the cleavage" between the sense perceived world and the scientifically conceived universe. Husserl's *Crisis* offers us an alternative account of it. In *The Crisis of European Sciences and Transcendental Philosophy* (1970), the cleavage is accounted for by showing how the universe depicted by natural science is an intellectual construct generated only at very high levels of abstraction, generalization, formalization, and idealization. These intellectual operations presuppose the pre-scientific lifeworld out of which they grow. The pre-scientific lifeworld thus functions as "the fundament of meaning" from which the sciences progressively abstract and, by abstracting from it, elaborate progressively more formalized, generalized, and idealized conceptions of reality (Husserl, 1970, 48-53; Gurwitsch, 1974).

Scientific conceptualizations and theorizing, therefore, are products of abstraction, generalization, formalization, and idealization. Higher levels of these intellectual operations are built out of lower levels of them; i.e., higher levels presuppose the mental capacity to perform the operations at lower levels (Husserl, 1970). Hence we can conceive of *a hierarchy of the sciences* in accordance with the level of abstraction, generalization, formalization, and idealization required for constructing each (Gurwitsch, 1974, 132-149). We shall here simply sketch such a hierarchy on a large canvas. Starting from the most concrete level and moving up in terms of *increasing* abstraction, generalization, formalization, and idealization, the hierarchy would be the following.

A) Pre-Scientific Level of Experience. At the most fundamental level is the *lifeworld* as experienced in pre-scientific ways, mainly through sense perception, practical action, and natural language (Schutz, 1962; 1973; Gurwitsch, 1974). The lifeworld is a particular sociocultural world. At this level objects are experienced as tools, decorative items, artworks, dwelling places, work places, places of worship, etc. Human beings are encountered in their social roles and as participants in social institutions and as governed by social norms and symbols. My own body and the bodies of other people are our bodies as we experience them in our ordinary, pre-scientific lives, what has been called "the lived body" (Merleau-Ponty, 1962). The lifeworld and the people and things in it are interpreted from within the common-sense perspective of the particular society whose reality it is (Gurwitsch, 1974, 3-32).

The language through which the lifeworld is interpreted is natural or ordinary language. This language exhibits a low level of abstraction and generalization. To say that it exhibits a "low level of abstraction and generalization" is to say that the meanings of natural language remain close to the pre-linguistic meanings from which they are derived, namely, the meanings that things and people are pre-linguistically perceived as having in our everyday dealings with them (Schutz, 1973; Gurwitsch, 1974).

B) First Level of Scientific Experience. This level of scientific conceptualization remains close to the pre-scientific lifeworld. It is the level we call the *social sciences* and the *humanities.* At this level the scientist attempts to interpret the multiple sociocultural lifeworlds in which different peoples live. People are therefore still conceptualized in terms of their social roles and as participants in social institutions and as governed by social norms. At this level the bodies of people are still their lived bodies, not the biological organisms of biological science. The various symbol systems of different cultures are interpreted in terms of the pre-scientific ways in which the people who inhabit those cultures interpret them (Schutz, 1962; Gurwitsch, 1974).

The social sciences and humanities are more abstract and generalizing than pre-scientific experience in the lifeworld: social scientists construct general claims and more or less abstract theories about societies, religions, cultures, economies, political systems, and the individuals and groups functioning within them (Schutz, 1962; Husserl, 1970; Gurwitsch, 1974, 113-149).

C) Second Level of Scientific Experience. At this level the scientist systematically abstracts from the social, religious, cultural, political, and economic meanings of things, living beings, and people. In other words, the various sociocultural interpretations of reality are systematically disregarded or set aside. In still other words, the scientist abstracts completely from the lifeworld and instead regards all realites whatsoever as biological, chemical, and physical things and processes. For example, the scientist systematically disregards the reality of the human body as a pre-scientifically lived body and instead conceptualizes this body exclusively as a biological organism, i.e., the human body as it is conceived in the biological sciences. This accordingly is the level of the *natural sciences* (Gurwitsch, 1974, 132-149)

High-level theories and concepts are developed in order to characterize these biological, chemical, and physical entities and events. By "high level" we mean theories and concepts that require extremely sophisticated systems of generalization, formalization, and idealization. An obvious example of such an extremely sophisticated system is advanced mathematics. One of the distinguishing characteristics of the natural sciences is their aim of depicting all things and events in mathematic terms and formulae.

Another example of a high level of idealization is the notion of strict causal determinism. From the pre-scientific experience of events in the lifeworld as occurring in a rough and ready causal fashion, one may begin to develop the notion of more precise and regular causal relations. As one imagines these relations being more and more precise and more and more regular, one can begin to imagine such precision and regularity approaching an ideal limit pole, namely, completely strict and exact causal relations for every event. By imagining this ideal limit pole one thus arrives at the notion of a strict causal determinism reigning throughout nature. One can then impose this idealized "cloak of ideas" on nature, and it can guide one's thinking and research as a natural scientist (Husserl, 1970, 48-53, 301-314, 353-378; Gurwitsch, 1974, 33-76).

VI. The Limits of Naturalism and "Naturalization"

Notice that naturalism finds it appropriate place *only at the second level of scientific experience.* Naturalism is incompatible with the pre-scientific level of experience in the lifeworld as well as the first level of scientific experience. The universe as interpreted by the natural sciences is the not the social world studied by the social scientist or humanities scholar nor is it the lifeworld as pre-scientifically experienced in ordinary life.

Let me give just one illustration of how the social sciences study a world composed of realities that can find no place in the universe that natural science explains. A thorough social inquiry would show that the lifeworlds of different peoples contain idealities that have no spatial or temporal locus (Gurwitsch, 1974, 92-94). Social norms and moral prohibitions, for example, are idealities, as are artworks and religious myths. In fact the meanings of natural language are also idealities. *To call something an "ideality" is to imply that it is omnitemporal and aspatial* (Gurwitsch, 1974, 64-66). Take, for instance, a religious myth. A myth can be told to several different groups of people at very different times and places and still remain the same myth. The myth can be interpreted in many different ways and still be the same myth. The myth is therefore not located in space or time. Of course, the myth, in order to be told, must be written or spoken in language. The sounds of the spoken words as well as the marks that compose the written words exist in space and time. But the same myth can be spoken at many different times and written in many different languages. Idealities, therefore, are part of the ontology of social lifeworlds (Husserl, 1969, 153-169).

Idealities, however, are not part of the ontology of the universe as explained by the natural sciences. In that universe everything must have a spatial location and a specifiable temporal duration. The natural scientist who works within the naturalistic attitude has already abstracted from idealities. He or she must systematically disregard all idealities in order to assume that it is necessary to specify the spatial locations and temporal durations of all objects and events. But idealities are only an example of the manifold properties of things, processes, and people from which one must abstract in order to think within the naturalistic attitude.

Naturalism is a systematic conception at which one arrives only at what we have called "the second level of scientific experience." Naturalism finds its valid application at that second level exclusively. The pre-scientific level of experience, the lifeworld, as well as the first level of scientific experience, the social sciences and humanities, cannot be "naturalized" without distorting them. They cannot be naturalized because naturalization requires an abstraction that would disregard the realities and idealities experienced at those levels. Naturalization systematically sets aside the properties of the lifeworld and the sociocultural worlds. Hence naturalization must remain confined to its own appropriate level, the second level of scientific experience, the natural sciences.

Each science thus conceptualizes one abstract dimension of the world. Now certain realities and idealities—in fact many of them—are composed of several of these dimensions. Living organisms, for example, are composed of physical, chemical, and biological entities and processes. Some living organisms also have psychological and even social features. Hence a *full* account of living organisms would require accounts of these different dimensions of their being. This full account, however, would include all the properties of the being, and hence the distinctive properties of each dimension would have to be included in it. It therefore makes no sense to speak of "reducing" one dimension to another although it does make sense to speak of concentrating on only one dimension and abstractively disregarding all others. If, however, one seeks a *full* account of a human being, one must include in it, for example, the cultural idealities of which he or she is aware and which accordingly inform his or her behavior. Naturalism would apply exclusively to the physical, chemical, and biological components of the human, not to the others.

VII. Phenomenology and Psychology also have their Genesis out of the Lifeworld

This phenomenological characterization of the genesis of the sciences allows us to return to our earlier discussion of the relationship between phenomenology and psychology and indicate their place in the hierarchy of sciences. The methods of both Husserlian phenomenology and empirical psychology are designed to generate high-level concepts through the operations of abstraction, generalization, and idealization. Phenomenology presupposes that its subject matter, mental life, can be "purified" of non-mental features so that "pure mental life" can be studied in its own exclusive right. Accordingly, phenomenology requires rather precise abstractions and generalizations through which mental life can be conceptually distinguished from non-mental realities and processes. In our prescientific experience in the lifeworld we do not clearly distinguish between our minds and our bodies. We do use phrases like "it's all in your mind" or "I can see it in my mind's eye," but grasping the meanings of these ordinary phrases requires no definite distinction between mind and body; and every action we perform involves a unity of body and mind that is so "natural" that we are heedless of it. Constituting the subject matter of Husserlian phenomenology, then, demands a conceptual separating out of different dimensions of reality that are pre-scientifically experienced as one.

Distinguishing between phenomenology and psychology, moreover, requires distinguishing between empirical (factual) features of mental life and eidetic features of it. The phenomenologist must abstract from or disregard the empirical features of the mental processes under study and focus solely on their essential ones. As we have seen, this abstracting is achieved through the method of free variation in imagination. The empirical psychologist has his or her own methods for abstracting from the complex factual givens of mental life and pinpointing the *general* factual ones. Through the systematic and rigorous application of these different sets of methods the different subject matters of the two sciences are constituted. Consequently, questions about the relationship of these two different sciences of the mind to one another can be answered only by reference to the differences in methods that generate the differences in focus and conceptualization.

Therefore, one of the tasks of a phenomenology of the sciences consists in tracing the genetic constitution of phenomenology itself as a philosophical science out of the lifeworld. If the lifeworld is "the fundament of meaning" for all the sciences, it functions as this always presupposed fundament for phenomenology too; and the "pure mental life" that phenomenology abstractly studies is the mind that each of us concretely experiences as fully integrated with his or her body (Husserl, 1970, 48-53). This self-reflexive dimension of phenomenology—that it must be able to account for itself—permeates phenomenological thinking and imposes on it a special kind of scientific responsibility.

References

Gurwitsch, Aron. 1964. *The Field of Consciousness,* Pittsburgh: Duquesne University Press, 1964.

—. 1966. *Studies in Phenomenology and Psychology,* Evanston: Northwestern University Press.

—. 1974. *Phenomenology and the Theory of Science,* edited by Lester Embree, Evanston: Northwestern University Press.

—. 1985. *Marginal Consciousness,* edited by Lester Embree, Athens, Ohio: Ohio University Press.

Husserl, Edmund. 1962. *Die Krisis der Europäishen Wissenschaften und die transzendentale Phänomenologie: Eine Einleitung in die phänomenologische Philosophie,* herausgegeben von Walter Biemel, Haag: Martinus Nijhoff.

—. 1970. *The Crisis of European Sciences and Transcendental Phenomenology: An Introduction to Phenomenological Philosophy,* translated by David Carr, Evanston: Northwestern University Press.

—. 1969. *Formal and Transcendental Logic,* translated by Dorion Cairns, The Hague: Martinus Nijhoff.
—. 1974. *Formale und transzendentale Logik,* herausgegeben von Paul Janssen, The Hague: Martinus Nijhoff.
Koehler, Wolfgang. 1971. "On Unnoticed Sensations and Errors of Judgment," in *The Selected Papers of Wolfgang Koehler,* edited by Mary Henle, New York: Liveright, pp. 13-39.
Merleau-Ponty, Maurice. 1962. *Phenomenology of Perception,* translated by Colin Smith, New York: The Humanities Press.
Schutz, Alfred. 1962. *Collected Papers, I, The Problem of Social Reality,* edited by Maurice Natanson, The Hague: Martinus Nijhoff.
Schutz, Alfred, and Thomas Luckmann. 1973. *The Structures of the Life-World,* translated by Richard M. Zaner and H. Tristram Engelhardt, Jr., Evanston: Northwestern University Press.

Contributions to Phenomenology

IN COOPERATION WITH

THE CENTER FOR ADVANCED RESEARCH IN PHENOMENOLOGY

---

1. F. Kersten: *Phenomenological Method.* Theory and Practice. 1989 ISBN 0-7923-0094-7
2. E. G. Ballard: *Philosophy and the Liberal Arts.* 1989 ISBN 0-7923-0241-9
3. H. A. Durfee and D.F.T. Rodier (eds.): *Phenomenology and Beyond.* The Self and Its Language. 1989 ISBN 0-7923-0511-6
4. J. J. Drummond: *Husserlian Intentionality and Non-Foundational Realism.* Noema and Object. 1990 ISBN 0-7923-0651-1
5. A. Gurwitsch: *Kants Theorie des Verstandes.* Herausgegeben von T.M. Seebohm. 1990 ISBN 0-7923-0696-1
6. D. Jervolino: *The Cogito and Hermeneutics.* The Question of the Subject in Ricœur. 1990 ISBN 0-7923-0824-7
7. B.P. Dauenhauer: *Elements of Responsible Politics.* 1991 ISBN 0-7923-1329-1
8. T.M. Seebohm, D. Føllesdal and J.N. Mohanty (eds.): *Phenomenology and the Formal Sciences.* 1991 ISBN 0-7923-1499-9
9. L. Hardy and L. Embree (eds.): *Phenomenology of Natural Science.* 1992 ISBN 0-7923-1541-3
10. J.J. Drummond and L. Embree (eds.): *The Phenomenology of the Noema.* 1992 ISBN 0-7923-1980-X
11. B. C. Hopkins: *Intentionality in Husserl and Heidegger.* The Problem of the Original Method and Phenomenon of Phenomenology. 1999 ISBN 0-7923-2074-3
12. P. Blosser, E. Shimomissé, L. Embree and H. Kojima (eds.): *Japanese and Western Phenomenology.* 1993 ISBN 0-7923-2075-1
13. F. M. Kirkland and P. D. Chattopadhyaya (eds.): *Phenomenology: East and West.* Essays in Honor of J. N. Mohanty. 1993 ISBN 0-7923-2087-5
14. E. Marbach: *Mental Representation and Consciousness.* Towards a Phenomenological Theory of Representation and Reference. 1993 ISBN 0-7923-2101-4
15. J.J. Kockelmans: *Ideas for a Hermeneutical Phenomenology of the Natural Sciences.* 1993 ISBN 0-7923-2364-5
16. M. Daniel and L. Embree (eds.): *Phenomenology of the Cultural Disciplines.* 1994 ISBN 0-7923-2792-6
17. T.J. Stapleton (ed.): *The Question of Hermeneutics.* Essays in Honor of Joseph J. Kockelmans. 1994 ISBN 0-7923-2911-2; Pb 0-7923-2964-3

Contributions to Phenomenology

IN COOPERATION WITH

THE CENTER FOR ADVANCED RESEARCH IN PHENOMENOLOGY

---

18. L. Embree, E. Behnke, D. Carr, J.C. Evans, J. Huertas-Jourda, J.J. Kockelmans, W.R. McKenna, A. Mickunas, J.N. Mohanty, T.M. Seebohm and R.M. Zaner (eds.): *Encyclopedia of Phenomenology*. 1997 ISBN 0-7923-2956-2
19. S.G. Crowell (ed.): *The Prism of the Self.* Philosophical Essays in Honor of Maurice Natanson. 1995 ISBN 0-7923-3546-5
20. W.R. McKenna and J.C. Evans (eds.): *Derrida and Phenomenology.* 1995 ISBN 0-7923-3730-1
21. S.B. Mallin: *Art Line Thought.* 1996 ISBN 0-7923-3774-3
22. R.D. Ellis: *Eros in a Narcissistic Culture*. An Analysis Anchored in the Life-World. 1996 ISBN 0-7923-3982-7
23. J.J. Drummond and J.G. Hart (eds.): *The Truthful and The Good*. Essays in Honor of Robert Sokolowski. 1996 ISBN 0-7923-4134-1
24. T. Nenon and L. Embree (eds.): *Issues in Husserl's* Ideas II. 1996 ISBN 0-7923-4216-X
25. J.C. Evans and R.S. Stufflebeam (eds.): *To Work at the Foundations.* Essays in Memory of Aron Gurwitsch. 1997 ISBN 0-7923-4317-4
26. B.C. Hopkins (ed.): *Husserl in Contemporary Context.* Prospects and Projects for Phenomenology. 1997 ISBN 0-7923-4469-3
27. M.C. Baseheart, S.C.N.: *Person in the World.* Introduction to the Philosophy of Edith Stein. 1997 ISBN 0-7923-4490-1
28. J.G. Hart and L. Embree (eds.): *Phenomenology of Values and Valuing.* 1997 ISBN 0-7923-4491-X
29. F. Kersten: *Galileo and the "Invention" of Opera.* A Study in the Phenomenology of Consciousness. 1997 ISBN 0-7923-4536-3
30. E. Ströker: *Husserlian Foundations of Science.* 1997 ISBN 0-7923-4743-9
31. L. Embree (ed.): *Alfred Schutz's "Sociological Aspect of Literature".* Construction and Complementary Essays. 1998 ISBN 0-7923-4847-8
32. M.C. Srajek: *In the Margins of Deconstruction.* Jewish Conceptions of Ethics in Emmanuel Levinas and Jacques Derrida. 1998 ISBN 0-7923-4953-9
33. N. Rotenstreich: *Synthesis and Intentional Objectivity.* On Kant and Husserl. 1998 ISBN 0-7923-4956-3
34. D. Zahavi (ed.): *Self-awareness, Temporality, and Alterity.* Central Topics in Phenomenology. 1998 ISBN 0-7923-5065-0

Contributions to Phenomenology

IN COOPERATION WITH

THE CENTER FOR ADVANCED RESEARCH IN PHENOMENOLOGY

---

35. R. Cristin: *Heidegger and Leibniz*. Reason and the Path. 1998 ISBN 0-7923-5137-1
36. B.C. Hopkins (ed.): *Phenomenology: Japanese and American Perspectives*. 1999 ISBN 0-7923-5336-6
37. L. Embree (ed.): *Schutzian Social Science*. 1999 ISBN 0-7923-6003-6
38. K. Thompson and L. Embree (eds.): *Phenomenology of the Political*. 2000 ISBN 0-7923-6163-6
39. O.K. Wiegand, R.J. Dostal, L. Embree, J.J. Kockelmans and J.N. Mohanty (eds.): *Phenomenology on Kant, German Idealism, Hermeneutics and Logic*. Philosophical Essays in Honor of Thomas M. Seebohm. 2000 ISBN 0-7923-6290-X
40. L. Fisher and L. Embree (eds.): *Feminist Phenomenology*. 2000 ISBN 0-7923-6580-1
41. J.B. Brough and L. Embree (eds.): *The Many Faces of Time*. 2000 ISBN 0-7923-6622-0
42. G.B. Madison: *The Politics of Postmodernity*. Essays in Applied Hermeneutics. 2000 ISBN 0-7923-6859-2
43. W. O'Brien and L. Embree (eds.): *The Existential Phenomenology of Simone de Beauvoir*. 2001 ISBN 0-7923-7064-3
44. F. Schalow: *Heidegger and the Quest for the Sacred*. From Thought to the Sanctuary of Faith. 2001 ISBN 1-4020-0036-7
45. T. Toadvine and L. Embree (eds.): *Merleau-Ponty's Reading of Husserl*. 2002 ISBN 1-4020-0469-9
46. J.J. Kockelmans: *Ideas for a Hermeneutic Phenomenology of the Natural Sciences*. Vol. 2: On the Importance of Methodical Hermeneutics for a Hermeneutic Phenomenology of the Natural Sciences. 2002 ISBN 1-4020-0650-0
47. J.J. Drummond and L. Embree (eds.): *Phenomenological Approaches to Moral Philosophy*. A Handbook. 2002 ISBN 1-4020-0770-1
48. D. Fisette (ed.): *Husserl's Logical Investigations Reconsidered*. 2003 ISBN 1-4020-1389-2
49. D. Zahavi, S. Heinämaa and H. Ruin (eds.): *Metaphysics, Facticity, Interpretation*. Phenomenology in the Nordic Countries. 2003 ISBN 1-4020-1754-5

Contributions to Phenomenology

IN COOPERATION WITH

THE CENTER FOR ADVANCED RESEARCH IN PHENOMENOLOGY

50. T.M. Seebohm: *Hermeneutics. Method and Methodology*. 2004
ISBN 1-4020-2617-X
51. D. Carr and C.-F. Cheung (eds.): *Space, Time, and Culture*. 2004
ISBN 1-4020-2823-7
52. L. Embree (ed.): *Gurwitsch's Relevancy for Congnitive Science*. 2004
ISBN 1-4020-2891-1

Further information about our publications on *Phenomenology* is available on request.

Kluwer Academic Publishers – Dordrecht / Boston / London